The Principles and Practice of
Photochemical Machining and Photoetching

A selection of photoetched parts produced in the UK, USA and Japan, showing (left to right): heat sink; lead frame (bottom); chopper disc (centre) and diaphragm plate (top); bonding tip (bottom); daisy-wheel printer (centre) and shadow mask (top); rotor (bottom), shaver foil (middle) and decorative grille (top).

# The Principles and Practice of Photochemical Machining and Photoetching

**D M Allen**

Cranfield Institute of Technology

**Adam Hilger, Bristol and Boston**

*British Library Cataloguing in Publication Data*

Allen, D. M.
 The principles and practice of photochemical
 machining and photoetching.
 1. Photoengraving
 I. Title
 686.2'327   TR970

 ISBN 0-85274-443-9


Consultant Editor: **A E de Barr**

Published under the Adam Hilger imprint by IOP Publishing Limited
Techno House, Redcliffe Way, Bristol BS1 6NX, England
PO Box 230, Accord, MA 02018, USA

Typeset by Mathematical Composition Setters Ltd, Salisbury, UK
Printed in Great Britain by J W Arrowsmith Ltd, Bristol

This book is dedicated to those who
suffered most during its preparation:
my wife, Ann, and sons, Richard,
Stephen and Andrew.

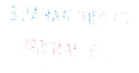

# Contents

# Preface

Although photochemical machining (PCM) has been practised as an industrial production technique for over twenty-five years and has been the subject of academic study for over ten years, there has never been a standard PCM reference text.

Whilst the excellent specialist books *Microphotography–Photography at Extreme Resolution* by G W W Stevens (London: Chapman and Hall, 1968) and *Photoresist–Processes and Materials* by W S DeForest (New York: McGraw-Hill, 1975) contain relevant material, no comprehensive text has been found covering design aspects, part processing, product diversity, technical considerations and economic implications of PCM.

By writing this book I hope that this void has been filled. The task would have proved impossible without the help of fellow members of the Photo Chemical Machining Institute and other industrialists who allowed me to visit their facilities in the UK, Western Europe, North America and Japan. I thank them most sincerely.

I also wish to thank Mrs Betty Warwick and Mrs Linda Nicholls for typing my manuscript.

Finally, I wish to acknowledge the help and encouragement given to me by Dr G W W Stevens and Dr D F Horne when I first started working in the field of photochemical machining at Cranfield Institute of Technology ten years ago.

**David M Allen**

# Introduction

## Historical background

The patterning of materials such as metal, glass and stone by wet chemical etching through apertures in adherent, etch-resistant stencils (also called resists or maskants) has been practised for hundreds of years. Originally, resists were applied in an appropriate pattern with the aid of a brush or they were applied overall and then scraped away from areas to be etched.

In the fifteenth century, a vinegar-based etchant was used in conjunction with a linseed-oil paint acting as a maskant to decorate iron plate armour. Other resists were developed from waxes, resins and other natural products such that, a century later, the technique was being used for intaglio print-making by etching into iron or copper plates through a wax 'ground' cut with a needle-point. It was not until the nineteenth century (after the discovery of hydrofluoric acid) that etching was used for the decoration of glassware.

The nineteenth century also saw the beginnings of photography as we know it today and with it came the development of a new type of resist known as a photoresist. It is applied overall and converted to a stencil by photomechanical processing. The coating is exposed to ultraviolet light through a high contrast photographic image to effect a photochemical reaction in the photoresist. As a result, the solubility of the reaction products differs from that of the starting material. This solubility difference is then exploited in the so-called 'development' process, which is really only a solvent wash, to form a stencil. (The processing is considered in detail in Chapter 3.) J N Niépce has been credited with the first photoetching, having succeeded in 1826 in etching pewter through a photoresist stencil formed from bitumen of Judea asphalt and developed in a mixture of lavender oil and mineral spirit.

In his British Patent 565 of 1852 William Fox Talbot described a photo-etching process for etching copper with ferric chloride solution through a

1

photoresist stencil made from bichromated gelatin developed in water. This was a very practical system and proved to be the forerunner of the method of manufacturing printed circuit boards.

Later, the etching of surfaces was complemented by the etching of perforations completely through materials. US Patent 378 423 assigned to John Baynes in 1888 describes the etching of materials from two sides through similar and dissimilar registered photoresist stencils. It therefore became possible to form recesses and perforations simultaneously.

In the early twentieth century improved photoresist formulations were developed by polymer chemists whose products, such as poly(vinyl alcohol), had reproducible batch-to-batch characteristics (unlike shellac, gum arabic and other natural products used in previous formulations) and greater resistance to the more aggressive, highly corrosive etchants. These materials needed to be sensitised by the addition of bichromate salts just prior to use and have a short shelf life once mixed together.

It was in the mid 1950s that the highly successful KPR family of photoresists was first marketed by Kodak. The formulation was based on pre-sensitised poly(vinyl cinnamate) and its appearance seems to coincide with the start of the photochemical machining (photoetching) industry.

Other 'firsts' in photoresist technology which have accelerated the growth of this industry include the introduction in the 1960s of Shipley AZ positive-working photoresists (based on the Kalle Kopierlacke formulations of the 1940s) and Du Pont Riston dry film photoresists. (For details of these systems see Chapter 3.)

Whilst photoresists are important elements in the photoetching process, it must not be forgotten that the equipment manufacturers have done much to convert a 'cottage industry' into a cost-effective, advanced manufacturing industry. This industry plays a valuable role world-wide in the production of precision parts and decorative goods (frontispiece) including colour TV shadow masks; integrated circuit lead frames; magnetic recording head laminations; heat ladders, plates and sinks; attenuators, light choppers and encoder discs; grilles, grids, sieves, meshes and screens; washers, shims and gaskets; jewellery; and decorative plaques and nameplates.

**The process**

The modern process of etching materials through photoresist stencils is shown in figure (I.1). Each process stage will be described in detail in this book under separate chapter headings (1 to 6) shown in the figure. The overall process is discussed in Chapter 7. The reader should also remember that whilst this book is concerned with photochemical machining and photoetching, similar etching processes outside the scope of this book are also used for the production of microelectronic devices from semiconductors and printed circuit boards. Etching must be acknowledged therefore as an extremely potent manufacturing process which has contributed greatly through developments in computing and information technology to the Second Industrial Revolution.

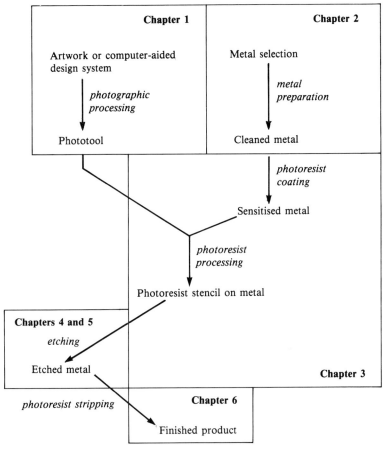

**Figure I.1** The photoetching process. Details of the various stages are found in the chapters shown in the layout.

## Terminology

As with all technologies that develop and expand rapidly over a short period of time in different parts of the world, terminology is difficult to standardise.

Photoetching is also known as
    photochemical machining (PCM) in the UK
    photo chemical machining (PCM) in the USA

photochemical milling
photomilling
photofabrication
chemical blanking (USA)
photoforming (USA, with an entirely different meaning in the UK; see §7.1.2)
and Etchineering® (Microponent Development Ltd, England).

The terms 'photochemical machining' and 'photoetching' are favoured by the author and in this text are intended to be synonymous. However, it has been suggested in past literature that photochemical machining implies substantial material removal and formation of perforations while photo-etching is a surface-etching technique concerned with the removal of relatively small amounts of material, hence the title of this book.

# Chapter 1
# Manufacture of Phototooling

# Chapter 1
# Manufacture of
# Phototooling

## 1.1 Introduction

In order to make components by PCM a phototool, usually comprising arrays of images on photographic film or plate (glass), must be manufactured (figure 1.1). These images may be produced by the more traditional method of photoreduction from enlarged 'cut and peel' artwork outlined in figure 1.2, or directly by using photoplotters or laser pattern generators developed recently for the production of printed circuit board phototooling.

The basic equipment, photographic techniques and materials used will now be described, followed by a brief review of artwork design requirements.

## 1.2 Coordinatographs

A coordinatograph is an accurate draughting machine equipped with a scribing tool capable of moving from point to point along orthogonal $x$ and $y$ axes. These points may be defined as cartesian coordinates $(x, y)$ and if the coordinatograph is fitted with a rotary table, then polar coordinates $(r, \theta)$ may be used also.

The usual method of draughting is to scribe with a scalpel blade into a red, strippable plastic layer coated onto dimensionally stable, clear polyester (polyethylene terephthalate) base. This 'cut and peel' material is sold under trade-names such as Ulano Rubylith, Kimoto Strip Coat and Huntermask. Areas bounded by scribed lines may be stripped off the support with a lifting tool comprising a thin, flat, blunt blade. The red layer absorbs green and blue light, the required illuminant for exposing orthochromatic photographic emulsions, while the polyester base transmits this

**Figure 1.1** A typical phototool consisting of stepped and repeated images. (Courtesy of Kodak Ltd, Harrow.)

radiation. As the artwork will be illuminated with transmitted light on the copyboard of the reduction camera, it is a good policy to minimise the amount of stripping and thereby help to reduce light flare in the reduction lens. This is usually achieved by stripping out features corresponding to apertures and etch bands (§1.10).

As many industrial components are based on concentric circle geometries, it is recommended that for PCM work a coordinatograph with a rotating table facility is purchased. Coordinatographs of this type are manufactured by Aristo, Coradi and Mutoh, to name a few of the more popular brands found in the UK. Reading of dials and verniers can become tedious and errors may be made if many coordinates need to be read for the production of complex artwork. To prevent problems of this nature, digital readout systems may be fitted, as shown in figure 1.3.

The plotting accuracy of a manual rotary coordinatograph (Aristo 4438)

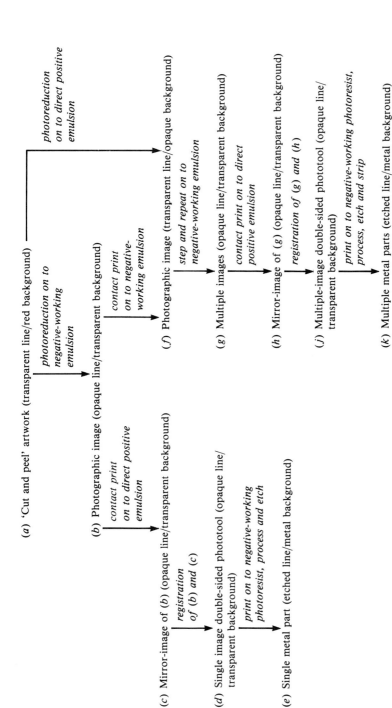

**Figure 1.2** Stages in the manufacture of parts from 'cut and peel' artwork together with tonality requirements.

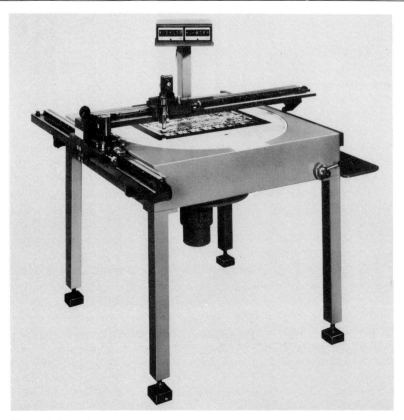

**Figure 1.3** Aristo coordinatograph 8738 with digital readout system. (Photograph courtesy of J Buck and Co, Cardigan.)

and a small circle device are shown in table 1.1. In this case accurate dimensions are obtained with the aid of movable scale tapes along the axes of the machine and recording dials, equipped with verniers, on the movable carriage holding the scalpel blade.

Other useful accessories for a manual coordinatograph besides the small circle device (maximum radius 10 mm) include a spotting microscope for

**Table 1.1** Plotting accuracy of the Aristo 4438 Coordinatograph.

|  | Plotting accuracy | |
| Basic equipment | Linear | Angular |
| --- | --- | --- |
| Aristo 4438 | ± 0.02 mm | ± 10″ of arc |
| Small circle device | ± 0.015 mm | − |

marking individual coordinates with a needle-prick, a beam compass for drawing arcs, and hand scribing tools used to draw irregular shapes.

The magnification factor for the artwork ($M$) is set by the tolerance of the photographic image according to the relationship

$$\text{Tolerance of the photographic image} = \frac{\text{Artwork tolerance}}{M}$$

$$= \frac{\pm 0.02 \text{ mm}}{M} \text{ for the Aristo 4438, for instance.}$$

So if the tightest tolerance of the photographic image is required to be $\pm 0.001$ mm, $M \geqslant 20$; or if $\pm 0.01$ mm, $M \geqslant 2$. When choosing a magnification factor, it must also be realised that:

(i) The artwork dimensions cannot exceed the work area of the coordinatograph.

(ii) The smallest feature of the artwork should not be less than 0.4 mm wide for ease of stripping.

(iii) The magnification factor ($M$) must equal the reduction factor ($R$) on the reduction camera, the range of which is often limited not only by the geometry of the camera but also by the specification of the lens.

The latest development in coordinatograph scribing has been brought about by the all-pervading influence of the computer. CAAG (computer-aided artwork generation) is capable of producing excellent results, as shown in figure 1.4. The stripping (peeling) is, of course, still done manually.

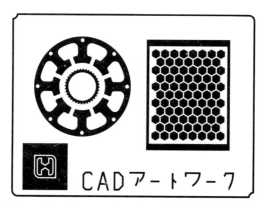

**Figure 1.4** Example of 'cut and peel' artwork produced on a Pentax CAD (computer-aided design) system. (Courtesy of Hirai Seimitsu Corporation, Osaka, Japan.)

A Wild–Heerbrugg (CADART) system is shown in figure 1.5. It comprises a computer, VDU, plotting table and optical digitiser. The scribing head has an acceleration of $5.0 \text{ m s}^{-2}$ and thus very quickly reaches its maximum speed of 300 mm s$^{-1}$. Automatic generation of vectors, symbols, arcs and circles, electronic step and repeat facilities and 'user friendly' technology make this type of system particularly attractive for PCM artwork generation.

**Figure 1.5**  A Wild–Heerbrugg CADART system used for CAAG (computer-aided artwork generation) comprising digitiser (right), computer and VDU (centre) and plotting table (left). (Photograph courtesy of Wild–Heerbrugg Instruments Inc, New York.)

## 1.3  Reduction cameras

Reduction cameras currently used in PCM applications appear to have been derived from graphic arts process cameras in their layout and construction, but features of cameras utilised in extreme resolution photography (ERP) for the manufacture of microelectronic masks have also been incorporated into the design to obtain the sharply defined, high contrast, high density images required for phototooling.

The quality of the reduced image is dependent on:

(i)   the quality of the camera construction and design,

(ii)  the optical quality of the reduction lens,

(iii) the photographic emulsion.

### 1.3.1  Camera construction

A reduction camera comprises a lens situated between the copyboard, on which the artwork is fixed by tapes, electrostatic charge or vacuum suction, and the photographic film (or plate) holder. The copyboard and film holder are necessarily parallel to each other and must also be perpendicular to the optical axis of the lens. If these conditions are not met, the recorded image will not be precisely focused at all points, as the image plane and emulsion plane will not be coincident.

**Figure 1.6**  The T231 camera, set up for photoreduction of opaque originals as might be used in decorative photoetching applications. This camera has a $1550 \times 1220 \, \text{mm}^2$ copyboard which can also be rear-illuminated. (Courtesy of DSR Littlejohn Ltd, Billericay.)

Most copyboards are vertical but some are horizontal, especially where the reduction factor is small, say less than 5. A large number of companies throughout the world manufacture a wide variety of reduction cameras (Figure 1.6) but they all have the following characteristics:

(*a*)  A source of uniform illumination; usually a number of fluorescent tubes, photoflood lamps or pulsed-xenon lamps. The complete emission spectrum may be used for rear illumination of 'cut and peel' film or for incident lighting of opaque artwork comprising black tape or Indian ink on white card, but frequently the light is filtered though a green band-pass filter. As this green illumination scatters less than blue light in silver halide emulsions, it produces a sharper image and, situated in the middle of the visible spectrum, is compatible with process lenses designed for colour work.

(b) Rigidity of construction. Some cameras are dedicated to one specific reduction factor, so that the copyboard-to-lens distance ($u$) and film-to-lens distance ($v$), together with the lens of fixed focal length ($F$) can be optimised for performance and then left permanently in place. In order to vary the reduction factor ($R$), it is necessary to be able to adjust both $u$ and $v$. This is accomplished by movements of the copyboard, lens and film holder along rails or lead screws.

It can be shown that

$$\Delta u = -RF\frac{\Delta x}{x}$$

$$\Delta v = \frac{F\,\Delta x}{Rx}$$

$$\frac{\Delta u}{\Delta v} = -R^2$$

where $\Delta u$ = the required change in copyboard-to-lens distance, $\Delta v$ = the required change in film-to-lens distance, $\Delta x$ = the change in image size required to go from the measured size to the desired size, and $x$ = the desired size.

(c) An exposure timing device or shutter.

(d) An accurate focusing mechanism. The focusing of the image is usually determined by allowing it to fall on a clear or ground glass plate and examining it with a microscope. Green light illumination is again favoured for this operation because the eye is most sensitive to this region of the visible spectrum. If any doubt exists as to whether optimum focus has been achieved, a photographic focusing test must be made by making a series of exposures on to the film with the focus altered by small increments (see also §1.7.3). When the images are processed, an operator can examine them at leisure and observe how quality first improves and then diminishes as focus passes through an optimum.

Many of the modern reduction cameras now include a calculator for determining the optimum positioning of lens, copyboard and film holder. Although shock-absorber mountings help to damp vibrations, which will affect image quality, it is advisable to keep a reduction camera at ground level where it can be 'bedded down'. In addition, a camera environment in which temperature, humidity and dust may be controlled is also recommended for precision work.

## 1.3.2  Optics

Process lenses are intended for working with the whole of the visible spectrum and are designed so that chromatic aberrations are reduced. Diffraction-limited lenses, on the other hand, are designed for use with specific monochromatic radiation (i.e. light of a single wavelength) and such a lens, although costly, will produce a better quality image if used with the ideal illumination source. The lens must have a flat field and pincushion and barrel distortions in the image must be completely absent.

## 1.3.3  Photographic materials

The conventional photographic materials used in PCM comprise a light-sensitive emulsion of silver halide grains in a gelatin matrix coated on to a backing material known as a support. In increasing order of dimensional stability the supports include cellulose triacetate, polyethylene terephthalate (often referred to as polyester) and glass (table 1.2). Optionally, the front or back surface of the support can be coated with a light-absorbing layer of dye to reduce halation.

**Table 1.2**   Dependence of phototool stability on emulsion support.

| Emulsion support | Thermal coefficient for size change (% per 1 $^{\circ}$C) | Humidity coefficient of linear expansion (% per % RH) |
|---|---|---|
| Cellulose triacetate | 0.0055 | 0.005–0.010 |
| Polyethylene terephthalate (polyester) | 0.0019 | 0.002–0.004 |
| Glass | 0.0010 | 0 |

The exposure that the emulsion receives is the product of the light intensity and the duration of the illumination. Exposure produces a silver atom as a speck on the silver halide grain, but as this latent image is invisible to the eye an amplification of the image is needed to render it visible. This amplification is known as development and entails bathing the latent image in an alkaline chemical solution (the developer) under safe-light conditions. Silver halide grains which contain silver specks are converted, by chemical reduction, to silver grains, whilst unexposed grains remain largely unaffected by this treatment.

As there is no end-point to development, emulsions are developed at a fixed temperature (usually 20 $^{\circ}$C) for a pre-set time. After development the film is stopped from developing further by washing in an acidic stop-bath, or sometimes this process is omitted and the film is just washed briefly in water prior to fixing.

In fixing, a process which desensitises the film, any remaining silver halide grains are washed out from the gelatin by dissolving them in a sodium or ammonium thiosulphate solution. The film is then washed to remove the fixer from the gelatin and dried. The resultant image may then be examined in ambient light or under the microscope.

A quantitative description of the photographic system may be obtained from an analysis of the characteristic curve obtained under a specific development process. By exposing the photographic material through a step-wedge (a strip of film consisting of density steps with approximately equal density increments between them) the characteristic curve (figure 1.7) may be drawn which can be used to measure contrast ($\gamma$) and maximum density ($D_{max}$).

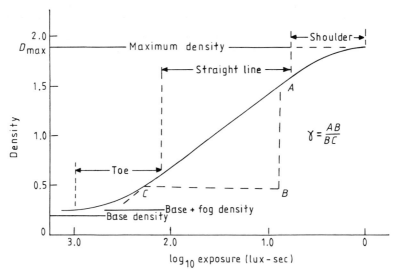

**Figure 1.7** Typical sensitometric curve for negative film with customary names of various parts of the curve.

Table 1.3 illustrates the differences between a pictorial 'black and white' film as used by a pictorial photographer and a variety of other emulsions used in photomechanical reproduction.

Line films can be developed in high contrast lith developers but conventional developers give increased exposure latitude, less corner rounding and less sharp edges. Direct-positive versions allow one to process positive-to-positive or negative-to-negative. Line films cost about a quarter of the cost of high resolution films. Although lith films produce very sharply defined edges, different line widths develop at different rates and determination of the best exposure time is difficult. Both lith and line films may be machine processed, this process being highly recommended for lith films.

For exceptionally fine precision work, high resolution emulsion (§1.7.1) may be used, although it is costly. If tonality needs to be maintained, then this emulsion is reversal processed. A suitable processing sequence and solution formulae are shown in table 1.4.

## 1.4   Step and repeat cameras

It is desirable from an economic consideration to etch as many components as possible from a single sheet of metal at one time. The reduction camera usually yields a single component image, and so to obtain an array of these images (figure 1.1) a contact step and repeat camera is used.

Two popular numerically controlled step and repeat cameras used in the UK are the Swedish Misomex 101 and Japanese Dainippon Superstep. Their specifications are listed in table 1.5 together with those of a small manual camera manufactured in England.

The single image is always in the form of clear lines on a black background. This enables any alignment marks or surplus lines to be masked off so that they do not get reproduced on to the image array formed after step and repeat. In addition, when the single image is printed on to negative-working emulsion, the pattern array will comprise black lines on a clear background, the tonality required for printing on to negative-working photoresist. Figure 1.2 shows how the tonality is changed during the manufacture of the phototool from artwork made on a coordinatograph.

## 1.5   Photoplotters

Photoplotters produce phototools by exposing photographic film or plate with a beam of light, the shape, size, intensity and movement of which is controlled by a computer. Prominent manufacturers of these machines include AEG, Aristo, Applicon, Calma, Computervision, Contraves, Ferranti, Gerber, Kongsberg, Sci-Tex and Quest Automation.

A typical CAD photoplotter system consists of: (i) a digitiser and graphic display(s); (ii) a computer to store and manipulate data and transform it to a form (floppy disc, paper or magnetic tape) suitable for input via its control unit to (iii) the photoplotter. The photoplotter comprises an attenuated light source and a film or plate holder which can be programmed to move relative to one another.

**Table 1.3** Emulsion characteristics.

| Emulsion | Pictorial black and white roll film | Line film | Lith film | High resolution | Bright light film |
| --- | --- | --- | --- | --- | --- |
| Grain diameter | ~ 1.1 $\mu$m | | | 0.05 $\mu$m | |
| Granularity | Very fine | | | Ultrafine | |
| Resolution | 69–95 lines/mm | | | > 2000 lines/mm | |
| Speed | Fast | Slow | Slow | Very slow | Very slow |
| Spectral absorption | Panchromatic | Orthochromatic | Orthochromatic | Orthochromatic | Ultraviolet light (340–380 nm) |
| Safe light | None | Dark red | Dark red | Dark red | White or gold fluorescent tubes |
| Base (support) | Cellulose acetate | Polyester | Polyester | Polyester (film) Glass (plate) | Polyester |

| γ (contrast) | Medium (γ = 0.5–0.7) | High (γ > 3) | Very high (γ = 5–15) | High (γ > 3) | High |
|---|---|---|---|---|---|
| $D_{max}$ | 1–2 | > 3 | > 3 | > 3 |  |
| Miscellaneous | | (1) Direct positive versions available. (2) Although development can be made with 'lith' developers, better results (increased exposure latitude, less corner rounding but slightly less sharp edges) are achieved with 'non-lith' developers. (3) Cost only a quarter of HR film. | (1) Different line widths develop at different rates (2) Determination of best exposure time difficult. | (1) Can be reversal processed | (1) Can be processed in lith and line developers. Particularly suited for rapid high temperature development (rapid access processing). |

19

**Table 1.4**  Reversal processing of high resolution emulsion (Kodak formulations).

| Stage | Process solution | Time |
|---|---|---|
| Develop | D-8 | 3 min |
| Wash | Water | 1 min |
| Bleach | R21a† $(1 + 9)$ | 2 min |
| Wash | Water | 30 s |
| Clear | 10% $Na_2SO_3$ (sodium sulphite) | 30 s |
| Expose to full room light in the above sulphite solution | | 15 s |
| Redevelop | D-8 | 2 min |
| Fix | Rapid fixer | 30 s |
| Wash | Water | 2 min |

†R21a comprises water (1 l), potassium dichromate (50 g) and concentrated sulphuric acid (50 ml).

**Table 1.5**  Specifications of some step and repeat cameras.

|  | Misomex 101 | Dainippon Superstep | Repro 1629 |
|---|---|---|---|
| Maximum original size | $230 \times 305$ mm$^2$ ($9'' \times 12''$) | $254 \times 305$ mm$^2$ ($10'' \times 12''$) | $169 \times 290$ mm$^2$ ($6\frac{1}{2}'' \times 11\frac{1}{2}''$) |
| Maximum area of stepped and repeated images | $762 \times 1016$ mm$^2$ ($30'' \times 40''$) | $1118 \times 1118$ mm$^2$ ($44'' \times 44''$) | $525 \times 795$ mm$^2$ ($20\frac{1}{2}'' \times 31\frac{1}{4}''$) |
| Tolerance per step | 0.005 mm (0.0002$''$) | 0.010 mm (0.0004$''$) | 0.020 mm (0.0008$''$) |

Designs are based on:

(*a*) A photohead moving in $X$ and $Y$ directions over a flat holder—suitable for films and plate.

(*b*) A stationary photohead with a flat holder moving in $X$ and $Y$ directions (again suitable for films and plate).

(*c*) A photohead capable of moving in a line parallel to the axis of a rotating drum carrying photographic film.

The attenuation of the light source is achieved by passing the beam through aperture plates or masks to alter its shape and size, and interposing a neutral density filter to alter intensity. It must be remembered that as plotting speed increases, light intensity must also increase in order to achieve uniform exposure over the whole plot.

All attenuations and movements of the light beam are carried out under computer control.

Faster (projection speed) negative-working lith and line emulsions are used in photoplotters. To ensure the best results from film products, it is recommended that machine processing be employed rather than manual dish processing. The former method ensures greater control over film-size changes as well as developer activity.

Lith emulsions produce the highest possible contrast and edge definition,

**Table 1.6** Photoplotter specifications.

|  | Quest EMMA 80 |  | Gerber 33B |  |
|---|---|---|---|---|
| Max. plot size (mm$^2$) | $581 \times 734$ | $(23'' \times 29'')$ | $635 \times 762$ | $(25'' \times 30'')$ |
| Average accuracy (worst case) |  |  | $\pm 0.0125$ mm | $(\pm 0.0005'')$ |
| (Static plot) | $\pm 0.015$ mm | $(\pm 0.0006'')$ |  |  |
| (Dynamic plot) | $\pm 0.035$ mm | $(\pm 0.0014'')$ |  |  |
| Repeatability | $\pm 0.006$ mm | $(\pm 0.00024'')$ | $\pm 0.0025$ mm | $(\pm 0.0001'')$ |
| Resolution | $0.0025$ mm | $(0.0001'')$ |  |  |
| Mask size (max.) | $5.08$ mm | $(0.20'')$ | $5.08$ mm | $(0.20'')$ |
| (min.) | $0.025$ mm | $(0.001'')$ | $0.050$ mm | $(0.002'')$ |
| Mask accuracy | $0.010$ mm | $(0.0004'')$ |  |  |
| Photoplotters speed (max.) | $80$ mm s$^{-1}$ | $(3.2''/\text{sec})$ | $127$ mm s$^{-1}$ | $(5.0''/\text{sec})$ |

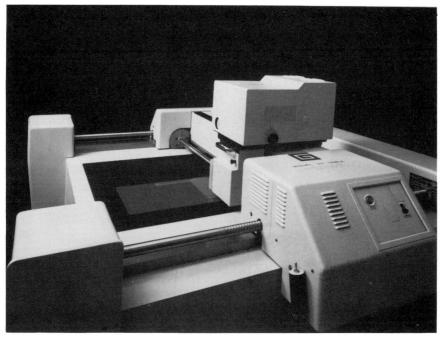

**Figure 1.8** Gerber 33 photoplotter. (Courtesy of the Gerber Scientific Instrument Company, Hartford, Connecticut.)

but are more difficult to handle than a line emulsion, which has wider exposure and development latitudes. Table 1.6 lists the specifications of two popular photoplotters. Both models are extensively used in printed circuit board manufacture and the production of phototooling for colour TV aperture masks (§7.3.1), encoder wheels, grids and other components made by PCM. Figure 1.8 illustrates the Gerber 33 photoplotter.

### 1.6  Laser pattern generators

These machines have been developed recently to speed up printed circuit board production. Typically, they can produce artwork 100 times faster than conventional photoplotters (§1.5).

In the case of the Excellon LPG 2000 (figure 1.9) a phototool is produced by arranging for a spot of laser light $(0.025 \times 0.025 \text{ mm}^2;$ $0.001 \times 0.001 \text{ in}^2)$ to scan over an area of photographic emulsion up to $457 \times 610 \text{ mm}^2 (18 \times 24 \text{ in}^2)$. Depending on whether the laser has been

**Figure 1.9** Excellon LPG 2000 Laser Pattern Generator. (Courtesy of Excellon Automation, Torrance, California.)

switched on or off, the spot will, or will not, expose the emulsion, with the 432 million ($18 \times 10^3 \times 24 \times 10^3$) decisions being controlled by a computer.

## 1.7 Extreme resolution photography (ERP)

ERP stretches microphotography to its limits and may be used in photo-etching and photoforming (§7.1.2), such processes being known collectively as microphotofabrication. The resultant products may be components, gratings, graticules, scales or microelectronic devices. ERP requires: high resolution emulsions; a perfect (aberration-free) diffraction-limited lens; and a precision focusing system.

### *1.7.1 High resolution emulsions*

Some of the characteristics of this type of emulsion are listed in table 1.3 but, in order to obtain ultrafine images—the object of the exercise—correct exposure and development are essential.

Overexposed plates result in image spread, which makes focusing (§1.7.3) difficult to judge. Exposure times in a $2^{1/4}$ time series such as 10, 12, 14, 17, 20, 24 . . . seconds will enable the correct exposure to be found. If any doubt exists, then underexposure is preferable.

A typical developer (such as Kodak D-8 or D-178) comprises hydro-quinone, alkali to activate it, sulphite as a preservative and bromide as an antifoggant (table 1.7). Not only does chemical development occur but physical development (a type of silver plating produced by absorption of silver salt from the unexposed grains followed by reduction to silver) increases the mass of exposed grains. As this process increases speed, density

**Table 1.7** Formulations of Kodak high contrast developers for use with Kodak HR emulsions.

| | D-8 | D-178 (stock solution) |
|---|---|---|
| Sodium sulphite (anhyd.) | 90 g | 90 g |
| Hydroquinone | 45 g | 45 g |
| Sodium hydroxide | 37.5 g | 18 g |
| Potassium bromide | 30 g | 30 g |
| Water to make | 1 l | 1 l |
| Dilution for use: | Undiluted | 2 vols. stock solution + 1 vol. water |
| Development time (20 °C) | 2–3 min | 3–4 min |

and contrast, development is strictly controlled by immersion for a specific time at a constant temperature to produce satisfactory results (Figure 1.10).

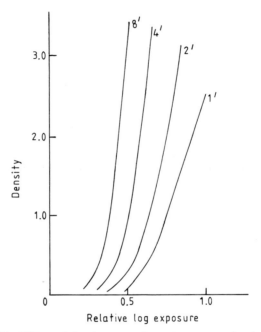

**Figure 1.10**   Effect of development time when processing in D-178 (2 parts + 1 part of water).

### 1.7.2   Diffraction-limited lenses

The resolution limit for a perfect lens ($R_{lens}$) can be found from the formula

$$R_{lens} = \frac{10^6}{\lambda(nm) \times f/number}$$

where $\lambda$ = wavelength of exposing light and $f$/number is the ratio of the focal length to the aperture. For 550 nm light and an $f/2$ lens,

$$R_{lens} = \frac{10^6}{1.1 \times 10^3} = 900 \text{ lines/mm}.$$

If this lens is used with HR emulsion capable of resolving 2000 lines/mm, then the overall resolution of the combined system $R_c$ is given by

$$R_c = \frac{R_{lens}\, R_{emulsion}}{(R_{lens}^2 + R_{emulsion}^2)^{1/2}}$$

$$R_c = 820 \text{ lines/mm}.$$

It can be seen that any improvement to overall system resolution is dependent on the use of difficult-to-manufacture, costly, low $f$/number (small focal length) lenses.

Microscope objectives are frequently used to form minute images but unfortunately they have a very limited field. The relationship of the focal length to the aperture in microscope objectives is described in terms of numerical aperture (NA) where

$$NA = 1/(2 \times f/\text{number}).$$

For a $\times 40$ objective with an NA of 0.65, then

$$f/\text{number} = 1/(2 \times NA) = 1/1.3 = 0.77.$$

A perfect (aberration-free) lens images a point source as a disc of light (Airy disc) with a radius ($r$) which is proportional to $\lambda$ and $f$/number and given by the equation

$$r(\mu m) = \frac{1.22 \times \lambda(nm) \times f/\text{number}}{10^3}$$

e.g. for $\lambda = 500$ nm and an $f/2$ lens, $r = 1.22$ $\mu$m. As a line may be considered as a succession of overlapping point sources, it gives rise to an intensity profile as shown in figure 1.11.

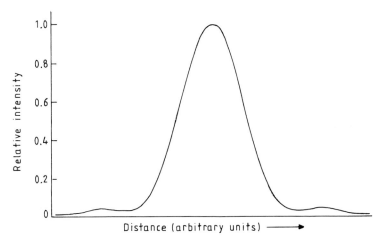

**Figure 1.11** Intensity profile of line image calculated from the overlapping of many Airy discs.

As the image of an 'infinitely fine' line possesses a finite width and unsharpness, this is the cause of a perfect lens having a resolution limit. When the centres of two parallel lines are brought close together their line profiles overlap. There is a critical separation of the two lines, below which there is no reduction in the light intensity between their centres, and resolution of the two lines is then physically impossible (figure 1.12).

The quality of line produced may be described quantitatively by a $z$ number (the ratio of resultant linewidth to reduction lens $f$/number). Thus, the profile quality of a 8.8 $\mu$m line produced with an $f$/5.6 lens aperture will be equal to that of a 4.4 $\mu$m line produced with an $f$/2.8 lens aperture ($z = 1.57$) and superior to that of a 4.4 $\mu$m line produced with an $f$/5.6 lens aperture ($z = 0.79$).

The resultant changes of line profile with width have the following consequences (Stevens 1968):

(i) since the relative intensity in the centre of an image of a line increases with its width, less exposure is needed for wider lines;

(ii) the width of the image of a line will widen as exposure is increased;

(iii) the narrower the line the more rapid will be the relative increases of width with increasing exposure.

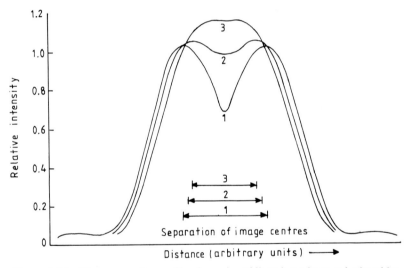

**Figure 1.12**   The composite profiles for pairs of lines have been calculated by adding the light intensity derived from each. Three cases are shown for separations indicated by the arrows. Note how rapidly the density depression between the images decreases with reduced separation (after Stevens 1968).

### 1.7.3   Focusing systems

It is necessary to position the image precisely on the surface of the HR emulsion. This may be achieved by:

(i) Observation of the image on the surface of a fixed-out, dried HR plate containing ink marks. The marks are used to fixate the focus of the eye when viewing through the back of the plate with a microscope.

(ii) Photographic focusing: altering the focusing position by known, repeatable increments (see §1.3.1 (*d*)). Further details of both methods are given in the paper by Allen, Horne and Stevens (§1.11).

For production of single small images, distance-pieces attached to microscope objectives may be pushed up against the emulsion but 'bruising' can result. A better system depends on the use of an air-gauge where a 1 $\mu$m distance change can be amplified to a 10 mm pressure change on a manometer—useful when imaging over large areas of HR plate where glass flatness is suspect.

## 1.8 Phototool materials

Although photographic images are produced on silver emulsion initially the phototool itself need not necessarily be made of this material. The most popular alternatives are film-based diazonium salt formulations (known as diazos), Image Plane Plate$^{TM}$† and chrome-on-glass.

Commercial silver emulsions have already been reviewed (§1.3.3) and show a very wide range of properties. Choice is not only limited to technical considerations, but also to cost and ease of processing.

Diazos are low cost alternatives to silver emulsions. They are made by contact printing from silver emulsion masters but appear to be used more in the USA than in the UK where diazo machine processing is not popular due to the nature of the chemicals used in development of the image formed after exposure to ultraviolet radiation.

Image Plane Plates are produced from high resolution silver emulsion plates using a patented ion exchange process to produce a permanent image, extending some 3 $\mu$m *into* the surface of a 1.5 mm thick soda lime glass sheet.

While all phototool images are necessarily opaque to ultraviolet light (actinic light for photoresists), diazos and Image Plane Plates are transparent in the visible spectrum—a great aid when attempting to register phototooling by optical techniques (§1.9).

Chrome-on-glass phototools are usually produced by sputtering chromium metal on to glass sheet, coating with photoresist, contact printing with ultraviolet light from a silver emulsion master, developing the resist image and etching the chromium film with a suitable etchant such as aqueous ceric ammonium nitrate solution.

† Trade mark of Precision Art Coordinators Inc, East Providence, Rhode Island, USA.

**Table 1.8** Phototool characteristics.

| Characteristics | Silver emulsion on film | Silver emulsion on glass (plate) | Diazo on film | Image Plane Plate (in glass) | Chrome on Glass |
|---|---|---|---|---|---|
| Support | Polyester | Glass | Polyester | Glass | Glass |
| Visual image | Opaque | Opaque | Transparent | Transparent | Opaque |
| Advantages | (1) Wide variety of emulsions to choose from to match technical and financial requirements | (1) Dimensional stability | (1) Low cost<br>(2) High resolution | (1) Dimensional stability<br>(2) Durability<br>(3) Damage can be repaired<br>(4) No surface coating means no parallax problems<br>(5) Chemically inert | (1) Dimensional stability<br>(2) Durability |
| Disadvantages | (1) Dimensional instability<br>(2) Easily damaged | (1) Identical emulsions more expensive on glass than on film<br>(2) Easily damaged<br>(3) Limited range of emulsions available | (1) Dimensional instability<br>(2) Easily damaged<br>(3) Needs to be developed in special machine | (1) Limited resolution (smallest feature 0.05 mm)<br>(2) Cannot be processed in-house | (1) Need to etch pattern into chromium film |
| Suitability for a PCM phototool | Short production runs | Short production runs | Short production runs | Long production runs | Long production runs |

Table 1.8 compares the characteristics of phototools made from the materials discussed.

## 1.9   Production of double-sided phototools and registration of images

In general, film-based materials are used as phototools, rather than glass plate, due to the lower cost of film products. However, as outlined in table 1.2, it can be seen that the major advantage of glass is its dimensional stability and where tolerances are small it may be necessary to employ glass phototooling.

Whilst a single-sided phototool can be used successfully with very thin metals and films, it is usual to use double-sided phototools in PCM to improve edge profile (§6.3). In this case the two images may be mirror images with the same tonality or they may be dissimilar with the same tonality if recesses are also to be etched into the metal at the same time.

As outlined in figure 1.2 the simplest method of producing mirror images is to contact copy an image (on film or plate) on to a direct positive emulsion. If fiduciary marks such as targets, bomb-sights or butterflies are incorporated into the original image, manual registration on a light table using optical aids such as microscopes, magnifying glasses or surgeon's loupes will be made easier. Once the exact superimposition of the two images has been achieved emulsion-to-emulsion then this relationship is maintained by pinning or bonding. Pinning of films is achieved by punching a round hole and one or more slots through the films on a registration punch and inserting precision stainless steel pins through them. The metal sheet to be etched may be punched or drilled (prior to coating with photoresist) so that the pins pass through this sheet also. Bonding of the two halves of the phototool may be achieved using liquid adhesives or pressure-sensitive adhesive tapes. To prevent distortion of the phototool, and hence misregister, when contact printing, a 'spacer' can be incorporated into the phototool. It should be equal in thickness to the photoresist-coated metal sheet to be imaged and comprises one or two thin strips of material attached along appropriate edges of the phototool.

It is possible when producing mirror-image double-sided film phototools to produce mirror images in exact register without fiduciary marks using a registration punch as follows:

(i) punch imaged film emulsion-side down;

(ii) under safe-light conditions punch a piece of direct positive film emulsion-side up;

(iii) under safe-light conditions pin the two films together emulsion-to-emulsion;

(iv) contact print on to the direct positive film;

(v) separate the films, process the exposed film and dry;

(vi) register the films together, emulsion-to-emulsion.

Both glass and film phototools can be registered by attaching them to registration frames with spray adhesive. The frames are equipped with two spring-loaded adjustment screws for altering relative positions in the $x$ and $y$ directions. The two halves of the phototool remain quite separate, with registration being carried out manually using optical techniques. Once registered the adjustment screws are locked in position.

Glass phototooling has also been registered using elastomeric hinges and pins set in drilled holes in the glass.

## 1.10  Artwork design

The usual practice in PCM is to make the phototools with dimensions which match those of the engineering drawing of the component *plus or minus* an etch allowance. This is a compensation for the undercut formed as a result of etching to an acceptable edge profile and is fully discussed in §6.4.2.

As the rate of etching is dependent on the line width in the resist stencil (§5.3.1) all features are surrounded by an etch band to ensure an even rate of etching on all edges. The width of the etch band should be equal to the width of the smallest feature corresponding to an aperture in the component, or approximately 0.8 mm if the smallest component aperture is wider than 1 mm. The etch band serves not only as a means of obtaining uniform profiles on all etched edges but also as a method of conserving etchant.

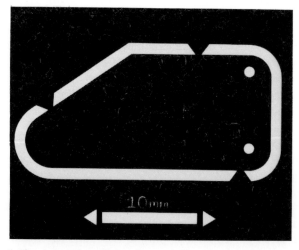

**Figure 1.13**  Artwork used for the manufacture of a spring steel shutter blade as described in §6.4.

To prevent components becoming detached from the main sheet and being lost in the etching machine sump and to prevent complex components becoming entangled when stripping off the resist, tabs are often drawn across the etch bands. These tabs are triangular, the apex pointing towards the component. After etching, the apex should be etched to a width of 0.1–0.13 mm (0.004″–0.005″) so that the components can easily be broken out from the retaining sheet (figure 1.13).

To aid registration of images and correct orientation of reduced images in the step and repeat camera, 'targets' or 'butterflies' (figure 1.14) are often incorporated into artwork, spaced as far apart as possible to increase accuracy of registration.

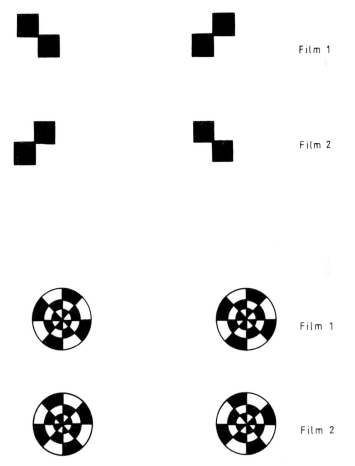

**Figure 1.14** Films 1 and 2 may be registered by superimposing 'butterflies' (top) or 'bomb-sights' (bottom) and checking alignment visually.

Angular features on artwork will be successfully reproduced on photo-tooling and resist stencils unless extreme reductions in size (>1000) are made (*SPSE Handbook*; section on microphotofabrication). However, etching of metal through the stencil will 'round off' these features to an amount dependent on the metal thickness (§7.1.1). If design specifications need these metal features to be 'sharpened up' then the artwork needs to be modified, typically by the addition of serifs to salient areas.

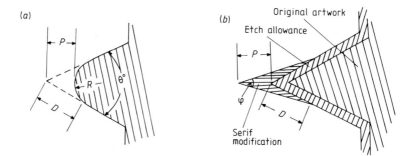

**Figure 1.15** (*a*) An angular feature of included angle $\theta$ in metal has been radiused off by the etching process. (*b*) The original artwork is shown with its etch allowance and serif modification. Note that no etch band is shown here.

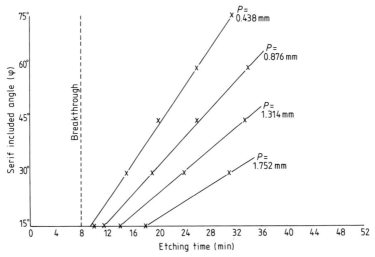

**Figure 1.16** Serif etching in stainless steel (thickness = 0.254 mm) showing etch times required for right angle formation.

The Photo Chemical Machining Institute Publication A-210 (1978) describes the serif shape quantitatively (figure 1.15) by the formulae:

$$P = R\left(\frac{1}{\sin{(\theta/2)}} - 1\right)$$

and

$$D = R\,\frac{1}{\tan(\theta/2)}$$

where $\theta$ = angle to be maintained and $R$ is the radius obtained with the original artwork.

If $R$ is dependent on the metal thickness $T$ (say $R = 0.75T$) and $\theta = 90°$ then

$$P = 0.75T(1.414 - 1) = 0.31T$$

and

$$D = 0.75T.$$

It is the author's experience that serif design is not unique—longer, narrower serifs being as effective as shorter, broader serifs (figure 1.16).

## 1.11  Bibliography

Allen D M, Horne D F and Stevens G W W 1980 Focusing of precision cameras for photofabrication mask production *J. Photogr. Sci.* **28** 136–9

Eastman Kodak Co 1977 Introduction to Photoplotting *Publication G-121* (New York: Eastman Kodak Co)

——1970 Techniques of Microphotography *Publication P-52* (New York: Eastman Kodak Co)

Photo Chemical Machining Institute 1978 Master Artwork Design for Photo Chemical Machining *Publication A-210* (PCMI, 4113 Barberry Drive, Lafayette Hill, PA 19444, USA)

——1984 Design and Artwork Generation Handbook *Publication H-1050* (PCMI, 4113 Barberry Drive, Lafayette Hill, PA 19444, USA)

Stevens G W W 1968 *Microphotography* 2nd edn (London: Chapman and Hall)

Thomas W Jr (ed) 1973 *SPSE Handbook of Photographic Science and Engineering* (New York: Wiley Interscience)

# Chapter 2
# Selection and Preparation of Metallic Materials for PCM

# Chapter 2
# Selection and Preparation of Metallic Materials for PCM

## 2.1 Material characteristics

### 2.1.1 Physical and chemical properties

The material chosen for part fabrication may be selected for its desirable physical properties such as high conductivity (copper), low specific gravity (aluminium), coefficient of thermal expansion (the alloy Kovar® has been developed such that its thermal expansion matches that of glass), high magnetic permeability (HyMu® 80), high melting point (tungsten), toughness (spring steel) or lustre (gold); or it may be chosen for its desirable chemical properties such as resistance to corrosion (titanium) or compatibility with secondary finishing processes (copper can be chemically blackened; aluminium can be anodised and dyed).

The ease with which the material can be etched (etchability) depends on its chemical composition, as etching is essentially a fast, controlled chemical corrosion reaction. Table 2.1 lists etchability ratings of some metals and alloys used in PCM and reflects the fact that pure metals have a wide range of reduction–oxidation (redox) potentials (see table 2.2) and corrosion resistance. Whilst it is relatively easy to chemically oxidise metals such as copper and nickel, it is difficult to oxidise refractory and noble metals and superalloys which have been developed especially to resist oxidation even at high temperatures.

### 2.1.2 Metal sheet stock specifications

The adage, 'You can't make a silk purse from a sow's ear' is relevant to PCM. To manufacture high quality products by PCM, the starting material

37

**Table 2.1** Etchability ratings of some selected metals and alloys used in PCM.

| Good | Good to Fair | Fair to Poor | Poor |
|---|---|---|---|
| Copper (rolled) | AISI 215 stainless steel | Molybdenum | Tungsten |
| Copper (electrolytic) | AISI 301 stainless steel | Nichrome (Ni, 20% Fe, 15% Cr) | Hastelloy C (Ni, 15% Mo, 14% Cr, 5% Fe, 3% W, 2.5% Co, 0.08% C) |
| Beryllium copper | AISI 302 stainless steel | Udimet alloys (e.g. Ni, 42% Fe, 12.5% Cr, 2.7% Ti) | Titanium |
| Brass (Cu, Zn) | AISI 304 stainless steel | Vanadium | Rene 41 (Ni, 19% Cr, 11% Co, 10% Mo, 3% Ti 1.5% Al) |
| OFHC copper | AISI 305 stainless steel | Chromium | Niobium (Colombium) |
| Phosphor bronze (Cu, 10% Sn, ⩽ 0.5% P) | AISI 316 stainless steel | Gold | |
| 90–10 copper/nickel | AISI 321 stainless steel | Lead | Tantalum |
| Zinc | AISI 347 stainless steel | Manganese | |
| Carbon steel | PH 15-7 stainless steel | Rhenium | |
| Kovar (54% Fe, 29% Ni, 17% Co) | PH 17-7 stainless steel | Zirconium | |
| Nickel | AISI 410 stainless steel | | |
| Monel (e.g. Ni, 31.5% Cu, 1.3% Fe) | AISI 420 stainless steel | | |
| Nickel silver (Cu, 25% Zn, 10% Ni) | AISI 430 stainless steel | | |
| Magnesium | Inconel alloys (Ni, 15% Cr, 7% Fe) | | |

*(continued)*

**Table 2.1** (*Continued*).

| Good | Good to Fair | Fair to Poor | Poor |
|---|---|---|---|
| Aluminium | Hastelloy B (Ni, 28% Mo, 5% Fe, 2.5% Co, 1% Cr, 0.5% V, 0.05% C) | | |
| Aluminium (anodised) | | | |

Courtesy of Photo Chemical Machining Institute.

needs to be 'photoetching quality'. This term implies that the material possesses the desirable properties required for compatibility with PCM processing; but, to date, these have never been defined. Fortunately, this situation is to be remedied by the Photo Chemical Machining Institute Metal Sheet Stock Specification due to be published in the very near future. The Draft Proposal covers, amongst other items, terms and definitions; acceptance and rejection criteria; and testing procedures. The purchaser of raw metal stock needs to specify chemical composition, temper and grain size, sheet/coil size, thickness (Appendix A) and other relevant data; but the

**Table 2.2** Standard reduction potentials of selected metals.

| | | | $E_0(V)$ |
|---|---|---|---|
| More noble metals | $Au^{3+} + 3e^- \rightleftharpoons Au$ | (gold) | 1.42 |
| | $Pt^{2+} + 2e^- \rightleftharpoons Pt$ | (platinum) | ~1.2 |
| | $Pd^{2+} + 2e^- \rightleftharpoons Pd$ | (palladium) | 0.83 |
| | $Ag^+ + e^- \rightleftharpoons Ag$ | (silver) | 0.7996 |
| | $Cu^{2+} + 2e^- \rightleftharpoons Cu$ | (copper) | 0.3402 |
| More reactive metals | $Pb^{2+} + 2e^- \rightleftharpoons Pb$ | (lead) | −0.126 |
| | $Sn^{2+} + 2e^- \rightleftharpoons Sn$ | (tin) | −0.1364 |
| | $Mo^{3+} + 3e^- \rightleftharpoons Mo$ | (molybdenum) | ~−0.2 |
| | $Ni^{2+} + 2e^- \rightleftharpoons Ni$ | (nickel) | −0.23 |
| | $Co^{2+} + 2e^- \rightleftharpoons Co$ | (cobalt) | −0.28 |
| | $Fe^{2+} + 2e^- \rightleftharpoons Fe$ | (iron) | −0.44 |
| | $Cr^{3+} + 3e^- \rightleftharpoons Cr$ | (chromium) | −0.74 |
| | $Zn^{2+} + 2e^- \rightleftharpoons Zn$ | (zinc) | −0.7628 |
| | $Ti^{2+} + 2e^- \rightleftharpoons Ti$ | (titanium) | −1.63 |
| | $Al^{3+} + 3e^- \rightleftharpoons Al$ | (aluminium) | −1.66 |
| | $Mg^{2+} + 2e^- \rightleftharpoons Mg$ | (magnesium) | −2.375 |

Specification will also be used as a consultative document for defining and quantifying defects such as burrs, discoloration, inclusions, pits, pocks, pores, projections, blisters, chatter marks, crown, dish, camber, lap, ripple, scale, twist, wavy edges and coil set.

The ideal metal for PCM, whether it be foil ($<0.125$ mm thick), strip ($<4.78$ mm thick, $<510$ mm wide) or sheet ($<4.78$ mm thick, $>510$ mm wide) should be flat, of uniform thickness (gauge) throughout, possess a fine grain size and uniform surface finish free from scratches, embedded particles and inclusions. The use of poor quality metal will result in substandard products as outlined in table 2.3.

**Table 2.3**  Effects of poor-quality metal on PCM.

| Metal defect | Effect on PCM process |
|---|---|
| Coil set (metal curved along its length) and dents | Difficult to coat with photoresist and to contact print against the phototool, resulting in loss of registration and detail |
| Gauge variation (e.g. crown, where the thickness increases from the edges to the centre of a strip of metal) | Difficult to determine optimum etching time as this is dependent on metal thickness |
| Too large a grain-size | Loss of resolution in etched features |
| Surface scratches | Difficult to coat with photoresist, with the result that etchant flows into the scratches and produces cosmetic defects on the surfaces of the product |
| Embedded particles and inclusions such as oxides, sulphides and silicates | Can produce defects on etching (e.g. pits, 'pimples' and loss of resolution in product) because they etch at different rates compared with the base material |

## 2.2  Metal preparation and equipment

In order to obtain good adhesion between metal and photoresist, the metal surface must be made perfectly clean by chemical cleaning (§2.2.1). Optionally, to improve adhesion further, the metal may be abraded (§2.2.2) or pre-etched (§2.2.3) to provide a rougher surface with a better 'key', or chemically treated to produce a surface layer of different chemical composition known as a conversion coating (§2.2.4).

After such treatments, described in more detail below, the material must be thoroughly rinsed with water and dried (§2.2.6).

### 2.2.1 Cleaning

The objective of cleaning the metal is to remove surface contamination, which occurs in the form of: oils, waxes, greases and other organic materials; rust, oxides and inorganic materials forming part of the surface; or dirt and particulate contamination.

It is difficult, within the scope of this text, to cover adequately the cleaning of all metals and alloys encountered in PCM. Proprietary solvent mixes, cleaners, deoxidisers and detergents proliferate throughout the world as metal cleaning is an important process not only in PCM but in all metal finishing industries (especially electroplating) and in printed circuit board manufacture.

Table 2.4 presents a general strategy for metal cleaning together with equipment requirements. Organic materials are removed first as they cover any inorganic contaminants. During this process dirt and particulate contamination are also removed to some extent. Electrocleaning is carried out by connecting the contaminated material as an anode (or for more efficient cleaning as a cathode) in an electrolytic cell containing, typically, a caustic cleaning solution. On passing DC current through the cell, gas evolution occurs at the electrodes. This gas 'scrubs' the surface clean.

The most common processing cycles appear to be: (i) alkaline or emulsifying soak cleaner to remove organic contamination; (ii) water rinse; (iii) acidic removal of inorganic contaminants; (iv) water rinse; and (v) drying. Such processes can be linked together in a conveyorised cleaning system as illustrated in figure 2.1. Processes (ii), (iv) and (v) are considered in §2.2.6. Apart from the solution chemistry, processes (i) and (iii) are identical in that the solutions are sprayed through two banks of vertical nozzles placed above and below the metal panel which is transported, by rollers, horizontally through the cleaning stations. If required, the cleaning action may be aided by incorporating non-abrasive nylon brushes into the cleaning station and flooding the surfaces of the panel with cleaning solution.

A good test for efficient cleaning is to spray water on the 'clean' metal surface and determine if the individual droplets will spread out to form a uniform, continuous thin film of water over the surface. This indicates adequate cleaning. Inadequate cleaning will produce uneven wetting of the surface. It should be remembered that many alternative cleaning processes exist (as shown in §2.2.5) and that new formulations are being developed all the time by commercial chemical companies. It must also be borne in mind that the economics of cleaning involve not only purchasing chemicals,

**Table 2.4** General strategy for metal cleaning.

| Contaminant | Cleaners | Equipment requirements |
|---|---|---|
| Oils, waxes, greases and other organic materials | (1) Chlorinated hydrocarbons (e.g. 1,1,1-trichloroethane) most effectively used in vapour degreasing equipment followed by | Tanks<br>Standard vapour degreasing plant |
| | (2) Alkaline soak cleaners which saponify fatty acid esters or the milder | Tanks or (more effective) conveyorised spray or brush cleaning system |
| | (3) Emulsifying soak cleaners containing anionic or non-ionic wetting agents, detergents, 'builders' (such as sodium metasilicate which provide a source of alkalinity, buffering and improve rinsability) and chelating agents (which soften hard water). | |
| Light organic contamination | (4) Acid emulsifying soak cleaners containing cationic wetting agents and phosphate 'builders' | Tanks or conveyorised spray or brush cleaning systems |
| Organic and particulate | (5) Alkaline, neutral and acidic anodic or cathodic (most efficient) electrocleaners | Standard electroplating plant |
| Rust, oxides and inorganic contaminants | (6) Neutral emulsifying soak cleaners<br>(7) Acidic aqueous solutions which convert insoluble contaminants to aqueous-soluble products | Tanks or conveyorised spray or brush cleaning systems |

but also disposing of them in a way that complies with environmental pollu-
tion regulations and purchasing equipment for their safe handling in a
factory environment in compliance with 'Health and Safety' regulations.

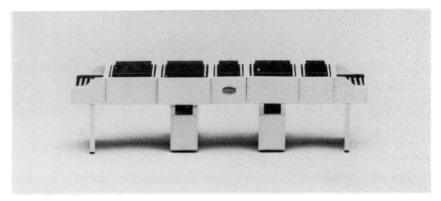

**Figure 2.1**   Conveyorised chemical cleaning system showing modular construction
of cleaning stations and rinse units. (Courtesy of CRH Electronics, Solingen, West
Germany.)

## 2.2.2   Surface abrasion treatments

If it is necessary for a new metal surface to be prepared, the original surface
can be removed physically with abrasives or chemically with an etchant
(§2.2.3).

Some abrasives (typically fine pumice) can be mixed with water to form
a slurry which may be sprayed on the metal surface. This system is used in
the Resco Jet Scrubber systems for preparing PCB copper foil surfaces.
Other systems favour passing the horizontal sheet metal flooded with slurry
through banks of vertical-spindle disc brushes or horizontal-spindle cylindrical brushes. Alternatively, water-lubricated abrasive brushes and discs or
3M Scotchbrite® can also produce excellent results.

It should be noted that the use of abrasives can cause problems to occur
at other stages of the PCM process. For instance, it is essential that every
last trace of abrasive is removed from the metal surface at the next rinsing
stage. If abrasive particles remain on the surface or, even worse, become
embedded in the metal then problems arise on coating with liquid
photoresists (when pinholes form), on etching the metal (when chemically
inert abrasives retard the etching of underlying metal), and after stripping
off the resist stencil (when the metal surface is found to be cosmetically
unattractive and/or unsuitable for electroplating or other finishing
processes).

### 2.2.3  Pre-etching

This process results in a fresh, rougher surface being formed on a metal sheet by etching away the smooth-rolled, original surface. The amount of material removed needs to be minimal ( $< 5 \mu$m per side). For these reasons pre-etching has also been termed 'key-etching' and 'micro-etching'.

The operation may be carried out using conventional PCM etchants (see table 4.1) or special formulations where etch rates are relatively slow, either by immersion in the etchant or by spraying the etchant on to the metal in an etching machine (§4.6).

The process is carried out when photoresist adhesion has to be exceptionally good and/or where the treatment gives other product design advantages. Pre-etching is most commonly found therefore in:

*TV shadow mask manufacture* (§7.3.1) where in addition to etching 300 000 perfectly formed, tapered slots it is required that the mild steel masks can be readily separated from each other when stacked. The pre-etched surface aids photoresist adhesion and serves to lower frictional forces between stacked masks.

*Magnetic recording head laminations* where the resist stencil acts as a bonding adhesive and adhesion to the metal must be exceptionally reliable (§3.6.6.).

*Articles made from difficult-to-etch materials,* such as titanium, where etching times may be prolonged or where the corrosive nature of an etchant may lead to chemical degradation of the resist stencil.

An obvious drawback to pre-etching is that matt, non-reflective surfaces are produced which may be cosmetically or functionally undesirable.

### 2.2.4  Conversion coatings

Conversion coatings are thin layers of metal salts or oxides formed on metal surfaces by reacting the metal with aqueous chemical solutions. These coatings not only alter the chemical structure of the surface and increase photoresist adhesion but also passivate the underlying metal. It has been suggested that conversion coatings improve etch factor (§5.2). As they are less soluble in etchants than the underlying metal, the rate of etching along the photoresist–substrate interface is reduced and this slower rate of lateral etching therefore produces larger etch factors as the rate of etching perpendicular to the surface remains unaffected.

The most commonly used solutions for conversion coating production have been dilute orthophosphoric acid to produce phosphate coatings on magnesium, nickel, nickel–iron alloys, carbon steels and zinc, dilute alkali-metal dichromate or chromic acid to produce chromate coatings on aluminium, copper and molybdenum, proprietary formulations for producing black oxide films on copper, and dilute nitric acid for increasing the thickness of passivating oxide films on stainless steels.

To produce sufficiently thick films, the aqueous solutions are used hot (70–80 °C) except with the reactive metals magnesium and zinc, when ambient temperatures are used. However, the coating must not be built up to a thickness such that it becomes powdery and only loosely bonded to the metal as adhesion will be impaired and etchant will then permeate rapidly along the photoresist–substrate interface.

It should be stressed that conversion coatings should only be used as a last resort in order to cure photoresist adhesion problems, and when deep etching is required.

Dissolution of conversion coatings with etchants formulated for etching the bulk underlying metal is not recommended. Breakthrough of the coating will occur at its weakest (thinnest) points first and a very rough etch will be obtained on shallow surface etching (see figure 2.2).

Ideally, before etching the bulk metal by conventional techniques, the conversion coating should be etched through the resist stencil apertures by immersion in an etchant which does not etch the underlying metal.

### 2.2.5  Recommended treatments compatible with photoresist coatings

The chemical condition of the metal after surface treatment must be compatible with the photoresist formulation (§3.5) to be coated on to the metal. A summary of required surface conditions is shown in table 2.5 together with specific cleaning cycle recommendations for four photoresists used in PCM.

### 2.2.6  Washing and drying

Washing and rinsing is required between processing stages to clean debris from the surface of the metal and also to stop contamination of processing baths by transference of drag-out from one bath to another. The water used for this purpose is usually sprayed on to the metal surface and, because large volumes of water are frequently used, is usually taken straight from the mains supply and used without further treatment. Distilled or deionised water can be produced at an additional cost and could be usefully employed in hard water districts for final rinsing but the use of water softeners and surfactants will also help to prevent deposition of salt residues on the metal after evaporation of surface rinse waters.

In order to conserve water, cascade-type systems have been developed which employ a number of successive rinse stations as shown in figure 2.3. The final rinse is carried out in fresh water sprayed from nozzle a, the rinse

**Figure 2.2**  Scanning electron micrographs (45° and ×210 (approx) horizontal magnification) of etched slots in (top) untreated stainless steel etched 0.028 mm deep and (bottom) strongly passivated stainless steel etched 0.027 mm deep.

**Table 2.5** Metal treatments for photoresist compatibility.

| Photoresist type | Recommended metal condition | Specific photoresist | Recommended cleaning cycle |
|---|---|---|---|
| Solvent and dry film negative-working photoresists | Acidic or neutral | Hunt Waycoat 450 | (1) Degrease<br>(2) Chemically clean with acidic solution *or* Chemically clean with alkaline solution and follow by acid dip |
| Positive-working photoresists | Preferably alkaline. Acidic or neutral acceptable | Shipley AZ 340 | (1) Carborundum BPA No. 1 Fine Buffing Powder<br>(2) Shipley Scrub Cleaner 11 or 70 *or* Shipley Neutra-Clean 68 or 7 |
| Aqueous dichromated fish gelatin | Acidic | Norland dichromated photo-engraving glue | (1) Abrasive clean with optional soak cleaning or electrolytic cleaning or pre-etching<br>(2) Acidic surface assured by dipping in 5% nitric acid if cleaning agent is alkaline. |
| Aqueous $Fe^{III}$ sensitised fish gelatin | Alkaline | Norland NPR 29F | (1) Hot alkaline soak detergent (e.g. MacDermid Metex 1726)<br>(2) Water rinse for 1 minute |

water collecting in sump A which overflows to sump B, whence it is pumped to nozzle b. This procedure is repeated at the various stations and, consequently, as the water becomes more polluted it is used at an earlier stage in the rinsing process.

Drying of the wet sheet metal stock can be accomplished by removing the bulk of water at ambient temperature using air knives or squeegee rollers with more thorough drying being carried out at elevated temperatures ($\geqslant 120\ ^\circ$C).

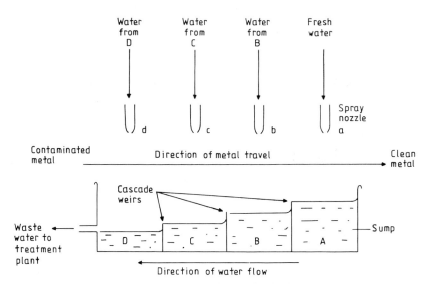

**Figure 2.3** Schematic of cascade rinsing system.

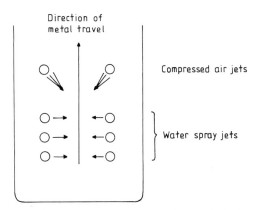

**Figure 2.4** Schematic of a combined rinsing and drying unit. The metal is pulled past the air knife after the water spray has been turned off.

This latter baking process also forms part of the dry film photoresist laminating process (§3.4). Although very thin material at ambient temperature may be passed through a laminator, thicker sheet metal will act as a heat sink, reducing the optimum photoresist flow temperature ($\sim 120\ ^{\circ}$C), so that inferior bonding is produced.

Air knives are simply jets of filtered, oil-free compressed air emitted through nozzles or apertures but orientated with respect to the metal sheet so that as the metal moves past the jets any surface water is blown in the opposite direction to that in which the sheet is travelling. This system may be incorporated in combined rinsing/drying units as shown in figure 2.4.

An alternative method of ambient temperature water removal is to pass the sheet metal through sets of squeegee rollers. Both methods are suitable for use in conveyorised processes.

Baking is most commonly carried out by placing racked metal sheets in conventional ovens, which may therefore be of considerable size. In conveyorised systems the metal may be passed through convection or infrared heating chambers, or more sophisticated units typified by the following:

A DEA turbine dryer which is designed to dry up to 6 mm thick stock up to 1 m wide by transporting it horizontally at 0–488 cm min$^{-1}$ on Tygon disc rollers, through vinyl squeegee rollers to the blast of hot air emanating from a 5 hp centrifugal turbine blower.

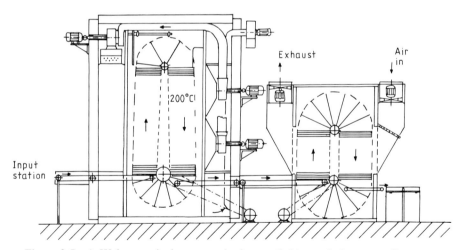

**Figure 2.5** A Weber vertical conveyorised oven (left) coupled to a cooling tower (centre) and sheet stacker (right). (Courtesy of Siegmund Inc, Hamden, Connecticut.)

A Weber vertical conveyorised oven (figure 2.5). This unusual configuration saves space and comprises pre-heat, heating and cooling zones together with a novel wire wicket transportation system.

## 2.3  Bibliography

Canning 1982 *Canning Handbook on Electroplating* 23rd edn (W Canning and Co
    Ltd, Great Hampton Street, Birmingham, England)
DeForest W S 1975 *Photoresist: Materials and Processes* (New York: McGraw-Hill)
    Chap. 3
Photo Chemical Machining Institute *Metal Sheet Stock Specification* (PCMI, 4113
    Barberry Drive, Lafayette Hill, PA 19444, USA) in preparation

# Chapter 3
# Modern Photoresist Technology

# Chapter 3
# Modern Photoresist Technology

## 3.1  Photoresist

Photoresist processes fit into PCM as outlined in Figure I.1. Photoresists serve two roles. Firstly, they provide a means of rendering substrates *photo-sensitive* and, secondly, they *resist* the etchant by providing a protective coating which adheres firmly to the substrate surface. This coating must not break down chemically or physically during etching of the substrate through apertures in the photoresist.

Whilst the latter role implies that thick coatings are required, the former role requires thin coatings in order to achieve good resolution. The correct coating thickness for any application is, therefore, the thickest one possible which is able to resolve the finest features in the photographic master image. In general, the average coating thickness is $2-5$ $\mu$m for liquid photoresists (§3.3) and $10-50$ $\mu$m for dry film photoresists (§3.4). Different coating methods (§3.3, §3.4) can be used to obtain uniformly thick, pinhole-free coatings. Uniformity of coating ensures consistent exposure characteristics at the printing stage of the process and absence of pinholes eliminates cosmetic defects in the substrate produced at the etching stage. Whilst the theory and concept of the photoresist are simple, many problems arise practically and these are discussed later (§3.7).

## 3.2  Negative- and positive-working photoresists

If a photoresist is negative-working, then exposed areas become insolu-bilised in the developing solvent. If it is positive-working, exposed areas become solubilised and will wash out in the developing solvent. These solubility changes are shown in figure 3.1, together with implications for subsequent etching processes. Advantages and disadvantages of the two systems are shown in table 3.1.

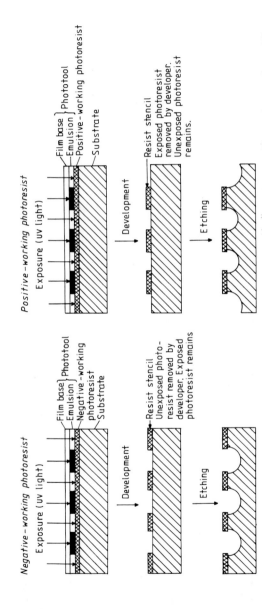

**Figure 3.1** Processing of substrates coated with negative- and positive-working photoresists. The phototool is of the same tonality in both cases.

**Table 3.1** Advantages and disadvantages of negative- and positive-working photoresists.

| Photoresist system | Advantages | Disadvantages |
|---|---|---|
| Liquid negative-working photoresist. | High resolution. Stencil usually tougher than that from positive resist, as it has been chemically altered during exposure. | Can 'heat fog', chip. More difficult to strip. Sensitive to oxygen. |
| Liquid positive-working photoresist. | Very high resolution due to intramolecular photolysis. Easily stripped (stencil composition identical to original photoresist composition). | Incompatible with alkaline etchants. Can be brittle and chipped easily. Needs careful, controlled processing. |

## 3.3   Liquid photoresist coating methods and equipment

Whereas dry film photoresists can only be applied to a substrate by laminating (§3.4) liquid photoresists can be applied by dipping, flowing, whirling, spinning, spraying and roller coating. However, in practice, dipping is used as the standard liquid coating method in PCM to the virtual exclusion of the other methods, which are only called upon for special applications.

### 3.3.1   Dipping

Dip-coating is achieved by the controlled withdrawal of a substrate through the meniscus of a liquid photoresist and as a consequence the two sides of a flat substrate are coated simultaneously with an equal thickness of resist. If the withdrawal is carried out at excessively high speeds, 'wedging' is likely to occur leading to variation in coating thickness over the metal sheet (§3.7.1). This coating method is very economical as small volumes of resist may be contained in a relatively narrow, but deep, tank with a close-fitting lid to prevent evaporation of solvent and minimise particulate contamination.

The mechanics of dip-coating comprise either raising of the substrate from a stationary tank, or lowering of the tank from a stationary substrate, or withdrawal of both tank and substrate from each other. All these methods for coating sheets may be found in industry together with modified techniques for reel-to-reel coating of coiled material.

The coating thickness is determined by the viscosity and temperature of the resist solution and the speed of withdrawal from the substrate. It is sometimes required that the viscosity of the commercial photoresist be reduced by addition of thinners (figure 3.2). To achieve identical coating thicknesses from day to day (and maintain constant exposure conditions) viscosity needs to be kept constant. A rapid quantitative assessment of photoresist viscosity is therefore required to monitor solvent evaporation effects.

The Zahn cup method of viscosity measurement is ideal for this purpose. The cups are robust, cheap, need only a small volume (44 ml) to fill them and can be used by non-skilled personnel with only two additional pieces of equipment—a stop-watch and a thermometer. A precisely dimensioned orifice is reamed into the bottom of the cup, its size depending on the viscosity of the fluid to be measured (table 3.2). The viscosity is determined by scooping a full cup of the resist from the reservoir and timing (in seconds) how long it takes to flow through the orifice, i.e. until the liquid stream breaks. The effects of temperature on viscosity can be quite pronounced so that a temperature versus viscosity curve for various resist concentrations should be consulted if coating-room temperatures vary from day to day. Resists used for dip-coating usually require a no. 1 Zahn cup. It should be noted that the viscosity is expressed in Zahn-seconds and not absolute units.

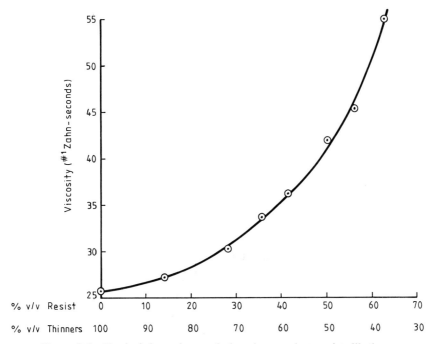

**Figure 3.2**    Typical dependence of viscosity on photoresist dilution.

An alternative method of measuring viscosity employs a Cannon–Fenske viscometer (figure 3.3) which is a modified form of an Ostwald (U-tube) viscometer. The Cannon–Fenske viscometer needs to be calibrated prior to use with a liquid (usually water) of known density and viscosity at the

**Table 3.2**    Zahn cup specifications.

| Zahn cup no. | Orifice size (in) | Viscosity range (cP) |
|---|---|---|
| 1 | 0.078 | 20–85 |
| 2 | 0.108 | 30–170 |
| 3 | 0.148 | 170–550 |
| 4 | 0.168 | 200–900 |
| 5 | 0.208 | 250–1200 and above |

temperature of the photoresist. The viscosity of the photoresist is determined by timing how long a column of the liquid takes to flow between two fixed points marked on a vertical length of capillary tubing under the influence of gravity.

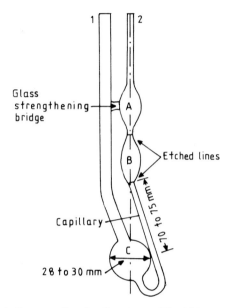

**Figure 3.3**   A Cannon–Fenske viscometer. Liquid is poured into reservoir C through 1. A column of it is pulled into reservoir A by applying suction to stem 2. On releasing the suction, the time taken for the liquid meniscus to travel between the etched lines is noted and used as a measure of viscosity by comparing with a known standard such as water.

The coverage of the resist may be found from the formula

$$\text{Coverage (sq. metres/litre)} = \frac{S \times D_l}{T \times D_s}$$

where $S$ is the solids content of the liquid resist expressed as weight percentage and $D_l$ is the density of the liquid resist (both found from manufacturers' literature), $T$ is the coating thickness (mm) and $D_s$ is the density of the dried photoresist layer. $D_s$ does not appear in the technical data of the majority of photoresists but it can be derived by weighing a layer of the dried resist in air and in water and applying Archimedes' principle (Allen *et al.* 1977).

### 3.3.2  Flowing

*Single-sided coating*

The photoresist is poured on to one side of the *horizontal* substrate which is then tilted at various angles so that the whole of the surface is covered. The coated material is allowed to drain by storing vertically and then dried. Some wedging may be present (see §3.7.1) and a uniform coating is difficult to achieve.

*Double-sided coating*

In continuous line processing, as used for the production of colour TV receiver tube aperture masks (see §7.3.1), the *vertical* metal strip can be flow-coated on two sides by dispensing the photoresist from nozzles situated at the top edge of the strip. Excess photoresist is allowed to drain off into a sump and is recirculated for economy. Due to the large volumes of photoresist required, it is not surprising to find the inexpensive, water-soluble formulations being used exclusively for this type of coating operation.

### 3.3.3  Whirling

The method may be used to coat one side of a metal sheet. The photoresist is poured on to the middle of the sheet and the 'pool' is then spread over the surface by rotating at low speeds (e.g. 50–100 rpm). The excess photoresist is flung off the edges of the sheet and the whirler therefore needs to be enclosed to contain this waste. Large whirlers with radii in excess of a metre have been used in the past to coat printing plates but smaller whirlers are useful for coating small metal sheets, the sheet being held on a rotating central plate by vacuum, or retaining lugs.

### 3.3.4  Spinning

This method is identical to whirling (§3.3.3) but substrate rotation occurs at high speed (500–5000 rpm) and with rapid acceleration such that the required rotational velocity is reached within a fraction of a second. This method produces very thin coatings and is used mainly in the microelectronics industry for the coating of oxidised silicon slices which are extremely flat and circular in shape.

### 3.3.5  Spraying

Although spraying is a rather wasteful process, requires a skilled operator, special air-driven or airless stainless steel spray guns and a spray booth, it is an excellent method for coating three-dimensional, or very large, parts.

An air-driven spray gun comprises either a pressurised container (pressure pot) or an open container (gravity bucket) filled with photoresist, powered by dry, oil-free compressed air at a pressure of approximately 20 psi. In an airless spray gun, air is replaced as the propellant by utilising the high vapour pressures of low-boiling-point solvents.

To obtain the best results, photoresist viscosity must be strictly controlled by additions of recommended thinners. Thin and discontinuous coatings imply the formulation is too dilute, while the appearance of 'cobweb strands' in the spray booth implies that the formulation needs thinning. If too much resist is applied to the substrate an 'orange peel' effect will be noted in the coating. It is also necessary that spray guns should be cleaned thoroughly after use to prevent nozzle clogging and contamination problems.

### 3.3.6  Roller coating

Roller coating and hot laminating (§3.4) are ideal coating methods to use when conveyorised processing systems are required. Due to the costly hardware required, roller coating is rarely found in the PCM industry but is used in making printed circuit boards where large quantities require streamlined and efficient processing.

## 3.4  Dry film photoresist coating and laminators

Dry film photoresist forms the middle layer of a three-layer sandwich structure (figure 3.4) supplied in rolls of various widths from 75–600 mm and in standard thicknesses, typically 0.010 mm for positive-working resist, 0.025, 0.037 and 0.050 mm for negative-working resist. The protective polyethylene layer is peeled off from the photoresist immediately before sticking the photoresist on to the hot substrate (§2.2.6). The coating is performed automatically by using a laminator (figure 3.4) containing a pressurised and heated roller transport system. Both sides of a sheet of metal can be coated simultaneously by using two separate rolls of photoresist.

After coating, a five-layer sandwich emerges from the laminator comprising polyethylene terephthalate, photoresist, metal, photoresist and

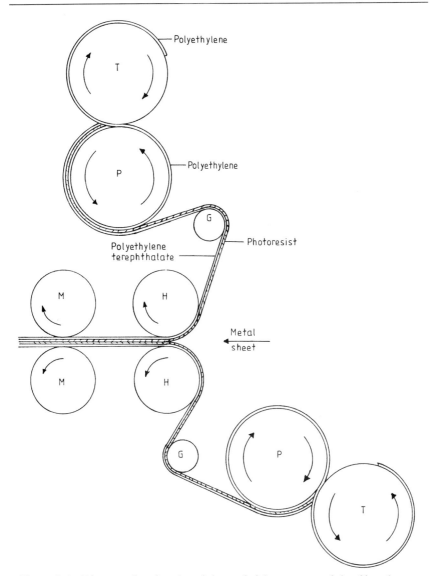

**Figure 3.4**  Diagram showing the triple sandwich structure of dry film photoresist together with the layout of a laminator. The laminator comprises two rolls of photoresist (P), two polyethylene cover sheet take-up rolls (T), guide rolls (G), heated, pressurised laminating rolls (H) and motor-driven rolls (M).

polyethylene terephthalate. The thin (0.025 mm thick) polyethylene terephthalate sheet protects the photoresist and on exposure through the phototool it is *not* usually removed. Keeping it in place prevents photoresist–photographic emulsion contact, with a subsequent loss of resolution due to light spread. If the protective sheet is peeled from the photoresist and contact is obtained by applying a vacuum between the surfaces, then a release coating of silicone wax sprayed onto the resist will enable separation after imaging or it may be dip-coated in a dilute poly(vinyl alcohol) solution and dried. The latter method is preferred, as silicone wax may inhibit development of the photoresist image. The advantages and disadvantages of using dry film photoresist are listed in table 3.3.

**Table 3.3**  Advantages and disadvantages of dry film photoresists compared with liquid photoresists.

| Advantages of dry film | Disadvantages of dry film |
|---|---|
| (1) No pinholing—reduced inspection and touch-up | (1) Lower resolution than liquid resists due to film thickness limitation. |
| (2) Can be used for 'tenting' (covering over holes). | (2) Cost. |
| (3) Tough and flexible. | (3) Restricted to use of specific photoresist thicknesses. |
| (4) No pre-exposure baking. | |
| (5) Fast coating time. | |
| (6) As no drying time is required contamination level is low. | |
| (7) Uniformity of coating. | |
| (8) Protective layer of polyester prevents damage and storage problems. | |

### 3.5   Photoresist chemistry

In choosing a photoresist for PCM process development it is useful to be able to test a range of photoresists with different chemical compositions for compatibility with the selected etchant. Commercial photoresists are sold under trade names which give no indication of their chemical constituents. Figure 3.5 gives a general classification of these photoresists according to chemical composition where known.

Summaries of the chemistry of each class of photoresist are presented in Appendix B. Further details may be found in the texts and papers listed in the bibliography (§3.8).

### 3.6   Photoresist processing and equipment

A general scheme of photoresist processing is outlined in table 3.4. As a rule, the recommendations of the photoresist manufacturer should be followed closely at all stages.

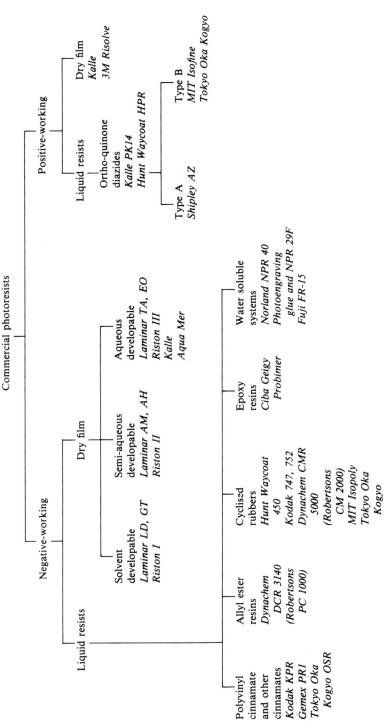

**Figure 3.5** Chemical classifications of commercial photoresists.

**Table 3.4**  Photoresist processing schemes.

| | Solvent developable, negative-working and aqueous developable, positive-working liquid photoresists. | Aqueous developable negative-working liquid photoresists. | Dry film photoresists. |
|---|---|---|---|
| 1 | Coat substrate | Coat substrate | Laminate onto substrate and trim |
| 2 | Dry | Dry | – |
| 3 | Pre-exposure bake | Pre-exposure bake | – |
| 4 | Expose | Expose | Expose |
| 5 | – | Chemical treatments† | – |
| 6 | Develop | Develop | Spray develop |
| 7 | – | Chemical treatments‡ | – |
| 8 | Water rinse | Water rinse | Water rinse |
| 9 | Dry | Dry | Dry |
| 10 | Touch-up | Touch-up | Touch-up |
| 11 | Post-development bake | Post-development bake | – |
| 12 | (Etch substrate) | (Etch substrate) | (Etch substrate) |
| 13 | Strip resist stencil | Strip resist stencil | Strip resist stencil |
| | *Options* To enhance visual contrast during inspection, the resist stencil may be dyed after stage 6 | *Options* Depending on the formulation used, the chemical treatments are optional, or involve sensitisation with $H_2O_2$(†) and/or hardening of the resist stencil (‡). Details are given in Appendix table B.1 | *Options* A post-development bake after stage 10. |

### 3.6.1  Pre-exposure baking

The recommended pre-exposure baking times and temperatures for liquid photoresists should not be exceeded. In negative-working photoresists, too high a baking temperature results in 'heat-fogging' (§3.7.2), whilst, in positive-working photoresists, high temperatures decompose the photoactive components in the coating with a resultant loss of photosensitivity.

The objective of pre-exposure baking is to drive off photoresist solvent from the liquid coating and thus pre-bake temperatures for solvent resists are typically 65–90 °C. Although baking times are kept as short as possible, the coating must be thoroughly dried by this treatment or it will adhere to the phototool during the vacuum exposure process (§3.6.2).

Heating equipment takes the form of conventional heating ovens (or heating chambers for continuous strip processing) or infrared heaters. Ideally the temperature should be constant in all parts of the oven or chamber to ensure uniform processing over the whole area of the metal sheet.

### 3.6.2 Exposure

The correct exposure $(E)$ for a photoresist coating depends on its thickness $(T)$, according to the equation:

$$E = e^{(aT+b)}$$

where $a$ and $b$ are constants. Over-exposure of a photoresist leads to difficulties in maintaining correct line widths and under-exposure leads to problems in being able to develop the image.

An essential aid for the determination of correct exposure is a neutral density step wedge. This narrow strip of photographic film comprises areas of different optical density, usually in increments of 0.15, as shown in figure 3.6. By exposing through this step wedge on to a substrate coated with a negative-working photoresist and developing the result, the correct exposure can be calculated. Ideally (for a 0.15 optical density step wedge) it is desired that the first 6 steps of the wedge pattern in the resist remain on the substrate after development, but that steps 7 and higher are partially or totally dissolved away. If step $N$ remains on the substrate after processing then the correct exposure time is found from the formula:

$$E = \sqrt{2}^{(6-N)} \times \text{test exposure time.}$$

The term $\sqrt{2}$ appears in the formula because 0.15 (the density step) $= \frac{1}{2} \times 0.30 = \frac{1}{2} \lg 2 = \lg \sqrt{2}$.

If a test exposure of 4 minutes results in 8 steps printed out on the substrate then the correct exposure

$$E = \sqrt{2}^{(6-8)} \times 4 \text{ min}$$

$$= \sqrt{2}^{-2} \times 4 \text{ min}$$

$$= 2 \text{ min.}$$

The correct exposure time for positive-working resists is found by exposing through the step wedge and calculating the smallest exposure which will completely solubilise the photoresist coating when developed. This is achieved by developing a deliberately over-exposed resist coating and noting the number of steps completely developed off the substrate. The percentage

**Figure 3.6**  Neutral density step wedge.

transmission of the corresponding neutral density step is then divided by
100 and multiplied by the original exposure time to give the correct exposure
time for that coating.

For example, if an exposure time of 2 min leads to 5 steps being completely developed off, then as the percentage transmission of step 5 = 22.4%
(see table 3.5), the optimum exposure = (22.4/100) × 2 min = 0.448 min =
26.9 s.

Exposure equipment used in PCM consists of a vacuum printing frame
and a source of high intensity ultraviolet light. Useful emission wavelengths
lie between the glass absorption cut-off region (approximately 280 nm) and
the blue region of the visible spectrum (approximately 500 nm). Ideally the
emission spectra should correspond to the maxima of the photoresist
absorption spectra. The most commonly used light sources include high
pressure mercury lamps, metal halide lamps, mercury xenon lamps, pulsed
xenon and actinic fluorescent tubes.

Commercial exposure units invariably contain two sets of light sources so
that both sides of coated material can be exposed simultaneously. By
monitoring light output, differences between two sources may be deter-

mined and exposures made identical by varying exposure times of the two sources.

As exposure $1 = I_1 t_1$ and exposure $2 = I_2 t_2$ where $I_1$, $I_2$ are intensities and $t_1$, $t_2$ are exposure times of sources 1 and 2 respectively, then for identical exposures to both sides of a photoresist coated material,

$$t_2 = (I_1/I_2) \times t_1.$$

$t_1$ is therefore set by the operator and $t_2$ is automatically calculated and effected. Ideally the optics of the printing unit should produce a collimated beam of ultraviolet light. This becomes increasingly difficult as the area of coverage (beam size) increases. Costs also increase.

Diffuse light sources (arrays of actinic fluorescent tubes) are relatively cheap and are found in the less expensive printers where twenty 40 watt tubes (ten per side) may be used to expose $585 \times 585$ mm$^2$ sheets.

**Table 3.5** Calculation of step wedge transmission.

| Step | Transmission optical density $= \log_{10}(100/\%$ transmission$)$ | % transmission |
|------|------------------------------------------------------|----------------|
| 1 | 0.05 | 89.0 |
| 2 | 0.20 | 63.2 |
| 3 | 0.35 | 44.8 |
| 4 | 0.50 | 31.6 |
| 5 | 0.65 | 22.4 |
| 6 | 0.80 | 15.9 |
| 7 | 0.95 | 11.2 |
| 8 | 1.10 | 7.8 |
| 9 | 1.25 | 7.5 |
| 10 | 1.40 | 7.24 |

Other sources are found in more expensive and sophisticated equipment. For example pairs of powerful (up to 4 kW) mercury vapour or metal halide lamps can be used to expose $660 \times 660$ mm$^2$ sheets on both sides with the advantage that exposure times are reduced considerably (figure 3.7).

Other exposure units commercially available are based on different designs such as:

(i) Those containing a single fixed ultraviolet light source, in which it is possible by means of a rotatable (e.g. flip-top) vacuum frame to expose each side of the coated sheet in turn. This system necessarily incurs the penalty of lengthening 'exposure time' by a factor of 2.

(ii) Those containing moving (or scanning) light sources. The vacuum printing frame is kept in a horizontal ($xy$) plane with two light sources (top

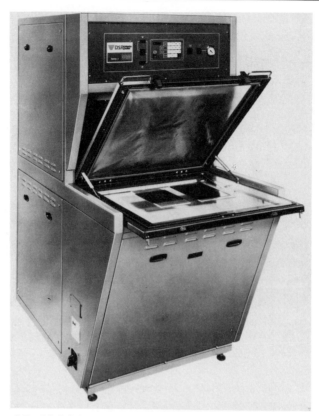

**Figure 3.7** Littlejohn 3000 Exposure Unit. Two different exposure programs may be pre-set if two different photoresists are being used. Two 2 kW metal halide lamps are used as light sources. (Courtesy of DSR Littlejohn Ltd, Billericay.)

and bottom) extending the width ($x$) of the frame. In operation, the light sources move over the length ($y$) of the frame, the speed being variable to effect different exposure times (figure 3.8).

(iii) Those containing stationary light sources and a mechanism which transports the phototool and its contents through the exposure system.

### 3.6.3  Development

Development of the photoresist coating consists in dissolving away either exposed or non-exposed areas of the photoresist, according to whether it is negative- or positive-working (§3.2). The developers are often proprietary

**Figure 3.8** Colight Scanex scanning exposure system fitted with two air-cooled mercury vapour lamps. (Courtesy of Astro Technology Ltd, Fareham.)

solutions formulated specifically for certain commercial photoresists (§3.5). As a general rule developers are used at $20 \pm 2\,^{\circ}C$, but some dry film photoresists are spray developed at higher temperatures. In particular, semi-aqueous developable dry film photoresists can be spray developed in aqueous sodium carbonate/organic solvent mixes at $30 \pm 2\,^{\circ}C$, whilst totally aqueous systems need to be spray-developed in sodium carbonate solution at $25–30\,^{\circ}C$. If these recommended temperatures are exceeded, solvent attack on the exposed areas of the photoresist will produce a matt, porous resist stencil.

Negative-working water-soluble photoresists are developed by spraying with tap water (table B.1). This is very economical and furthermore does not pose problems with regard to solvent flammability and effluent disposal.

It should also be noted that dilution of the alkaline aqueous developers used with positive-working photoresists affects the speed of development and the line width control (Elliott 1977).

Equipment utilised for development of photoresist varies considerably within the PCM industry. Simple immersion tanks may be used but faster, more efficient development is achieved by spraying developer on to the imaged photoresist. This applies particularly to dry film photoresists.

In order to achieve faster throughput of products, conveyorised spray developing machines are utilised which may be linked into systems comprising conveyorised rinsing and drying facilities also (figure 2.1).

### 3.6.4    Post-development baking

This process takes place after development, water rinsing and air-drying. Its purpose is to dry completely and toughen the stencil, particularly where solvent development causes photoresist swelling and solvent entrapment in the coating.

If recommended temperatures and baking times are exceeded, the stencil will become difficult to strip. In the case of dry film photoresists, post-development baking is optional but may be advantageous in deep etching applications. Equipment suitable for post-development baking is described in §3.6.1 (pre-exposure baking).

### 3.6.5    Stripping

The art of good stripping is to be able to remove the stencil completely from the metal substrate without staining or corroding the metal surface. A range of proprietary products is available for stripping the strongly adherent stencils obtained from negative-working solvent soluble photoresists. Some formulations, mostly used at room temperature, only soften and lift the coatings which then need to be brushed gently off the substrates. Other formulations, usually more toxic and requiring good ventilation facilities, are used hot and dissolve the stencils (table 3.6). Other stencils can be stripped relatively easily—a process consideration which should not be forgotten when choosing a photoresist system. After stripping, a brief 'acid dip' may be given to the etched metal parts, prior to water rinsing and drying. Equipment suitable for stripping resist stencils is described in §3.6.3 (development).

### 3.6.6    Bonding

As there is a limitation to the thickness of metal sheet that can be etched successfully, bonding methods can be used to laminate etched thin sheets together and thus build up thickness. This can be achieved by stripping the resist from the etched sheet, applying adhesive, laminating the sheets in register on a jig and heat-curing the assembly whilst keeping it under pressure. Although this method is very effective, it is labour intensive and time consuming. A more economical method is to use the resist stencil itself as an adhesive. In particular, cyclised rubber photoresists have been

**Table 3.6**  Photoresist stencil strippers.

| | Negative-working photoresists | | | Positive-working photoresists |
|---|---|---|---|---|
| | Liquid | | Dry film | Liquid and dry film |
| | *Solvent soluble* | *Water soluble* | | |
| | *Cyclised rubbers*: Hunt Waycoat CM, Indust-Ri-Chem J-100 and J110 NP strippers dissolve stencils at 100 °C, while Hunt Waycoat PF stripper lifts stencils at room temperature | *Norland Photoengraving Glue*: by immersion in 5–10% sodium hydroxide at 72–82 °C for 30–60 seconds or electro-lytically by connecting the resist-covered part as an anode in a 10% solution of sodium hydroxide containing 5–10% sodium hypochlorite (Ueda 1982) | *Solvent soluble*: Methylene chloride at 20 °C<br><br>*Semi-aqueous soluble*: 5% sodium hydroxide at 55 °C<br><br>*Aqueous soluble*: 5–10% potassium hydroxide at 35–55 °C | Many organic solvents including acetone, dimethylformamide and the Cellosolves (glycol ethers) at room temperature. Also strongly alkaline (sodium or potassium hydroxide) solutions |
| | *Polyvinyl cinnamates*: Chemical Processes 322 and 400 strippers lift stencils at room temperature | | | |

successfully utilised in this manner (Kodak 1972). If stronger bonding is required, a small percentage of an epoxy adhesive hardener can be added to the liquid photoresist prior to coating (Rembold 1978).

A very recent development in photoresist technology (and an obvious one in retrospect) is to use light-sensitive epoxy formulations as adhesive photoresists. The chemistry of these epoxy photoresists is discussed in Appendix B.1.

Such bonding methods are especially useful for manufacturing magnetic recording heads from special materials such as HyMu 80 and Vitrovac 6025 (see §7.3.5). The bonding of etched frets (arrays) of laminations is carried out by stacking the resist-coated frets in a press whilst maintaining the assembly at 135–230 °C in an oven or furnace. Each fret contains a registration system to ensure correct alignment and orientation of the individual laminations in the press. After bonding the heads are cut out from the laminated frets and the tabs ground off.

## 3.7   Photoresist problems

Whilst it should be realised that processing of photoresists is not difficult, occasionally problems can arise in trying to obtain the ideal photoresist coating or stencil. The most commonly encountered problems associated with the processing of negative-working solvent soluble photoresists may be detected by visual or microscopic observation of the coating at each stage of processing as outlined below.

### 3.7.1   Dip-coating defects

After pre-exposure baking of liquid resists, the coating should appear uniformly smooth and glossy and be of even coloration throughout. Observable defects include: streaks, caused by touching the wet coating against the sides of the photoresist container or by dipping attached clips or jigs or perforations in the sheet metal (used for holding material or in achieving registration), into the meniscus of the photoresist; and pinholes and entrapped dust, caused by using particulate-contaminated photoresist or insufficient metal cleaning.

Another defect can arise if dip-coating is carried out at excessively high speed ( > 250 mm min$^{-1}$). Known as 'wedging', the coating will be thinner at the top and thicker at the bottom of the metal sheet. This means that the correct exposure times for these two areas will be different and that any overall exposure of the coating will necessarily not be ideal for all parts of the coating. The remedies for curing the above defects are obvious.

### 3.7.2 Defects caused by incorrect pre-exposure baking

Too rapid drying of thick coatings can lead to reticulation of the photoresist film. This effect, also known as 'orange-peel', is due to uneven liberation of solvents from the wet film and solvent entrapment once a 'skin' forms on the film surface. Photosensitivity of the coating is thereby reduced, development will partially dissolve coatings which should have been photo-insolubilised and any stencil formed may allow permeation of the etchant through it.

In negative-working photoresists, too high a baking temperature may result in 'heat fogging'—a residue of polymerised photoresist in areas that have not been exposed to ultraviolet light (figure 3.9). The residue comprises thermally polymerised photoresist and its occurrence may be prevented by lowering the baking temperature or omitting pre-bake altogether and allowing the coating to air-dry at room temperature under safe-light conditions for several hours. This defect, together with others described in §§3.7.3 and 3.7.4, will of course only be detected after exposure, development and post-development baking of the coating.

**Figure 3.9** Photomicrographs ( × 70) of developed resist stencils. The stencil on the left is heat fogged.

### 3.7.3 Defects caused by incorrect exposure of the photoresist and/or poor quality phototooling

In addition to achieving the correct exposure of the photoresist with the aid of the step wedge (or photometer) as outlined in §3.6.2, it is also a prerequisite that phototooling be kept in intimate contact with the photoresist by means of a vacuum to effect dimensionally accurate image transfer. Additionally a vacuum also prevents oxygen from acting as a polymerisation inhibitor so that cross-linking of polymers to the desired photoproduct is not impaired.

If the phototool lacks density in its image then transmission of actinic light through the 'black' areas may cause sufficient cross-linking of the coating to make a 'scum' remain on the metal after resist development. This may be especially noticeable if exposure times are relatively prolonged.

### 3.7.4  Defects caused by incorrect development and rinsing

Development should be carried out ideally for the recommended time in the recommended, fresh developer, but in a production environment developers are used until they approach exhaustion, i.e. when it becomes apparent that development time is becoming excessively long. However, insufficient development in used developer, too short development in fresh developer or insufficient post-development water rinsing may result in residues being left on what should be a clean metal surface. This will cause retardation of etching and loss of resolution in etched features.

### 3.7.5  Stencil chipping

Some photoresist formulations (e.g. some of those based on poly(vinyl cinnamate)) result in the formation of a rather brittle stencil which may be

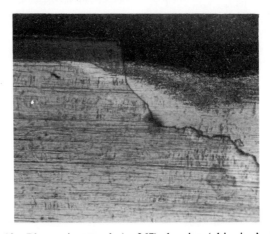

**Figure 3.10**  Photomicrograph ( × 267) showing 'chipping' of a resist stencil. From left to right is shown the undercut flap of photoresist, a fractured edge and extra undercut after the flap broke off, to the detriment of the edge profile.

'chipped' by an etchant spray when etching has produced undercutting of the stencil (see figure 3.10). Once chipping has occurred then edge definition will be lost on further etching. Other photoresists, such as those based on cyclised rubbers, are much more flexible and durable.

## 3.8   Bibliography

Allen D M, Horne D F and Stevens G W W 1977 Properties of liquid photoresists used in the photoetching of stainless steel *J. Photogr. Sci.* **25** 250–3

DeForest W S 1975 *Photoresist: Materials and Processes* (New York: McGraw-Hill)

Elliott D J 1977 Increasing the functional speed of positive photoresist *SPIE, Semiconductor Microlithography II* **100** 48–56

Kodak Ltd 1972 A new process for chemically milled parts *Kodak Photoresist Review* no. 23

Nakane H, Yokota A, Asaumi S and Hatazawa T 1973 The product control and the thermo-stability of positive photoresist *Photopolymers: Principles, Processes and Materials* 92–101

Photo Chemical Machining Institute 1978 Photoresists in photo chemical machining *Publication M-510* (PCMI, 4113 Barberry Drive, Lafayette Hill, PA 19444, USA)

Rembold K-H 1978 Polymerisable epoxy resins and their application in the field of printed circuits *Proc. 1st Printed Circuit World Convention, London, June 1978* pp. 1.11.1–1.11.7

Shimizu S and Bird G R 1977 Chemical mechanisms in photoresist systems: Part I— Photochemical cleavage of a bisazide system *J. Electrochem. Soc.* **124** 1394–1400

Steppan H, Buhr G and Vollmann H 1982 The resist technique—a chemical contribution to electronics *Angew. Chemie Int. Ed. Engl.* **21** 455–469

Technical literature, data sheets and handbooks of photoresist manufacturers, e.g. *Shipley Positive Photo Resist Photofabrication Technical Manual for Printed Circuits and Chem-Milling Applications* (April 1982).

Ueda Y 1982 A novel stripping process of water-base photoresist *PCMI Journal* no. 10 p. 8

# Chapter 4
# Etching, Etchants and Etching Machines

# Chapter 4
# Etching, Etchants and
# Etching Machines

## 4.1 Introduction

Etching comprises a heterogeneous chemical reaction in which liquid reacts with a solid substrate and oxidises it to produce a soluble reaction product. A wide variety of materials can be etched (see table 2.1) and a selection of etchants for the more commonly encountered sheet metals and foils are listed in table 4.1. In PCM the range of etchants is usually restricted to the less dangerous ones, namely aqueous ferric chloride solutions (often modified with additives), cupric chloride solutions, diluted mineral acids and some alkaline etchants based on sodium hydroxide or ammonium salts.

As can be seen from table 4.1, ferric chloride will attack a very wide range of metals, and consequently it has become the most widely used etchant in the PCM industry as a result of its versatility and low cost.

## 4.2 Mechanisms, kinetics and thermodynamics of etching

When a metal is immersed in an aqueous solution a potential difference is set up at the interface and, on the Helmholtz–Perrin model, an electrical double layer is formed consisting of two sheets of electrical charge, one on the metal surface and the other a fixed distance away in the liquid.

The latest models of the double layer are more complex, recognising the fact that water in contact with metal does not possess the same structure as water in the bulk of the liquid. In the proximity of the metal surface water molecules are preferentially orientated by their dipole moments.

To effect chemical attack on a metal surface a reactive, solvated ionic species must be transported through the Helmholtz layers to the metal where it must then oxidise it (and consequently be reduced itself). The ionic reaction product(s) must then be solvated and transported away from the

**Table 4.1**   Etchants for sheet metals and foils.

| Metal (composition) | Etchant formulation (v/v ratios where known, except where stated) | Temperature ($^\circ$C) |
|---|---|---|
| Alfenol (Fe, 16% Al) | (1) 42 $^\circ$Bé† FeCl$_3$ | 49 |
| | (2) conc. HNO$_3$ : conc. HCl : H$_2$O (1 : 1 : 2) | 38–49 |
| Aluminium | (1) conc. HCl : H$_2$O (1 : 4) | 22–65 |
| | (2) 20% NaOH solution | 60–90 |
| | (3) 12–20 $^\circ$Bé FeCl$_3$ | 49 |
| | (4) alkaline potassium ferri-cyanide solution, e.g. K$_3$Fe(CN)$_6$(329 g l$^{-1}$) : NaOH(16 g l$^{-1}$) : Na$_3$PO$_4 \cdot$ 12H$_2$O(30 g l$^{-1}$) | 55 |
| | (5) conc. HCl : conc. HNO$_3$ : H$_2$O(10 : 1 : 9) | 49 |
| | (6) 40 $^\circ$Bé FeCl$_3$ : conc. HCl (100 : 9) | 43 |
| Anodised aluminium | as for aluminium etchants (2) and (3) | |
| Beryllium‡ | ammonium bifluoride (NH$_4$HF$_2$) (90–180 g l$^{-1}$) | 27–32 |
| Beryllium copper | see copper alloys | – |
| Chromium | 42 $^\circ$Bé FeCl$_3$ : conc. HCl (2 : 1) | 32 |
| Columbium | see niobium | – |
| Constantan (55% Cu, 45% Ni) | 36–42 $^\circ$Bé FeCl$_3$ | 49 |
| Copper and copper alloys | (1) 30–42 $^\circ$Bé FeCl$_3$ | 43–49 |
| | (2) 33 $^\circ$Bé CuCl$_2$ (acidic) | 54 |
| | (3) modified chromic acid (e.g. Hunt Multi-Circuit Etch) | 27–38 |
| | (4) ammonium persulphate (220 g l$^{-1}$) | 32–49 |
| | (5) cupric ammonium chloride [Cu(NH$_3$)$_4$]Cl$_2$ (2M) | |
| | (6) hydrogen peroxide/sulphuric acid (e.g. Shipley Hydro-etch 536) | 45–50 |

*(continued)*

**Table 4.1** *(Continued)*

| Metal (composition) | Etchant formulation (v/v ratios where known, except where stated) | Temperature (°C) |
|---|---|---|
| Gold | (1) aqua regia i.e. conc. HCl : conc. $HNO_3$ (3 : 1) | 20–32 |
| | (2) potassium iodide (saturated) + iodine (20 g l$^{-1}$) | 50 |
| HyMu 80, 800 (80% Ni 4% Mo, rest Fe) | 42 °Bé $FeCl_3$ : conc. HCl (9 : 1) | 43–49 |
| Inconels (Ni, Cr, Fe) | 42 °Bé $FeCl_3$ | 54 |
| Kovar (Fe, 29% Ni, 17% Co) | 42 °Bé $FeCl_3$ | 49 |
| Lead | 33 °Bé $CuCl_2$ (acidic) | 54 |
| Magnesium | 10–20% conc. $HNO_3$ ( + additive such as Hunt X-flex to improve etch factor if required) | 35 |
| Molybdenum‡ | (1) alkaline potassium ferri-cyanide solution e.g. $K_3Fe(CN)_6$ (200 g l$^{-1}$) : NaOH (20 g l$^{-1}$) : sodium oxalate (5 g l$^{-1}$) | 55 |
| | (2) 40 °Bé $Fe(NO_3)_3$ | 40–55 |
| | (3) conc. $HNO_3$ : HF : $H_2O$ | |
| | (4) conc. $HNO_3$ : conc. $H_2SO_4$ : $H_2O$ (1 : 1 : 3) | 50–54 |
| | (5) conc. $HNO_3$ : conc. HCl : $H_2O$ (3 : 3 : 4) | |
| Moly Permalloy (Ni, 13% Fe, 5% Cu, 4% Mo) | 42 °Bé $FeCl_3$ : conc. HCl (9 : 1) | 54 |
| Monel (67% Ni, 33% Cu) | 42 °Bé $FeCl_3$ | 49–54 |
| Mumetal (Ni, 16% Fe, 5% Cu, 1½% Cr) | 42 °Bé $FeCl_3$ | 49–54 |
| Nickel | (1) 38–42 °Bé $FeCl_3$ | 49 |
| | (2) conc. $HNO_3$ : conc. HCl : $H_2O$ (1 : 1 : 3) | |

*(continued)*

**Table 4.1**   *(Continued)*

| Metal (composition) | Etchant formulation (v/v ratios where known, except where stated) | Temperature ($^\circ$C) |
|---|---|---|
| Nickel–iron alloys (e.g. Invar, Alloy 42) | 42 $^\circ$Bé $FeCl_3$ (with $HNO_3$ additions if necessary) | 49 |
| Nimonics ($\sim 80\%$ Ni, 20% Cr) | $FeCl_3 : HNO_3 : HCl$ | |
| Niobium‡ | conc. $HNO_3 : HF : H_2O$ (7 : 1 : 2) | |
| Phosphor bronze (Cu, 5% Sn, $\frac{1}{2}$%P) | 42 $^\circ$Bé $FeCl_3$ | 25–49 |
| Platinum | aqua regia, i.e. conc HCl : conc. $HNO_3$ (3 : 1) | 20–25 |
| Silver | (1) 36 $^\circ$Bé $Fe(NO_3)_3$ (2) 50–90% $HNO_3$ | 43–54 38–49 |
| Stainless steels | 35–48 $^\circ$Bé $FeCl_3$ | 35–55 |
| Stainless steels (containing molybdenum) | 36–42 $^\circ$Bé $FeCl_3$ with $HNO_3$ additions | 35–55 |
| Steels (mild, spring, silicon and tool) | 36–42 $^\circ$Bé $FeCl_3$ | 52 |
| Tin | 42 $^\circ$Bé $FeCl_3$ | 54 |
| Titanium‡ | (1) 10–50% HF (optionally with additions of $HNO_3$) (2) ammonium bifluoride ($NH_4HF_2$) : conc. HCl : $H_2O$ | 30–49 |
| Tungsten | alkaline potassium ferricyanide solution, e.g. $K_3Fe(CN)_6$(200 g $l^{-1}$) : NaOH (20 g $l^{-1}$): sodium oxalate (5 g $l^{-1}$) | 55 |
| Vanadium | conc. $HNO_3 : H_2O$ (1 : 1) | 25 |
| Zinc | 20–25% conc. $HNO_3$ ( + additives such as Hunt Rocket Etch to improve etch factor if required) | 36 |
| Zirconium‡ | (1) conc. $H_2SO_4$ : HF (2) conc. $HNO_3$ : HF : $H_2O$ | 36 36 |

†For definition of $^\circ$Bé see §4.3.2.
‡It should be noted that solutions containing HF or high concentrations of HCl will attack etching machine components made from titanium.

metal surface back into the bulk solution so that other reactive ions can be transported to the surface.

The rate of etching will therefore be controlled according to which of the three reactions, transport of the reactant, chemical reaction, or transport of the reaction product, is the slowest.

Little published work has been carried out in practical investigations of reaction kinetics, but those of Maynard *et al.* (1984a, b) are most enlightening. The reaction mechanism of the ferric hexaaquo ion (found in aqueous ferric perchlorate solution—an etchant *not* used commercially) with mild steel is thought to be as follows:

$$Fe(H_2O)_6^{3+} + Fe(H_2O)_x(s) \xrightarrow{\text{slow}} Fe(H_2O)_x \cdot Fe(H_2O)_6^{3+}$$

ferric hexaaquo   hydrated surface                          fast
ion            iron atom

$$\downarrow \text{fast}$$

$$Fe(H_2O)_x^+(ads) + Fe(H_2O)_6^{2+}$$

$Fe(H_2O)_6^{3+}$                         ferrous hexaaquo
$H_2O$     fast                    ion

$$2Fe(H_2O)_6^{2+}$$

ferrous hexaaquo ion

The simplified overall reaction is therefore:

$$Fe + 2Fe^{3+} \longrightarrow 3Fe^{2+}.$$

The reaction rate is limited by the rate of encounter of ferric hexaaquo ions with the steel surface. The rate of disappearance of ferric hexaaquo ions is dependent upon the first-order ferric hexaaquo ion concentration and the steel surface area and is inversely dependent upon solution volume and viscosity. This indicates diffusion controlled kinetics and the enthalpy of activation, $\Delta H = 2.8 \, \text{kcal mol}^{-1}$, is in the range for diffusion of small molecules in solvents of low viscosity. In addition, the reaction is neither inhibited nor catalysed by ferrous hexaaquo ions.

Mathematically, diffusion controlled kinetics are described by the equation:

$$-\frac{d[M]}{dt} = \frac{ADC}{S}$$

where $-d[M]/dt$ is the rate of dissolution of metal ($\text{mmol s}^{-1}$), $A$ is the surface area of metal exposed to the etchant ($\text{cm}^2$), $D$ is the diffusion coefficient ($\text{cm}^2 \, \text{s}^{-1}$), $C$ is the concentration of etchant ($\text{mol l}^{-1}$), and $S$ is the diffusion layer thickness (cm).

The mechanistic situation is more difficult to resolve when using ferric chloride solutions because complexes are formed between hydrated ferric ions and chloride ions. Complexed water molecules are displaced by chloride ions as the concentration of the solution (or concentration of chloride ion) is increased. The principal species in solution appear to be $FeCl_2(H_2O)_4^+$ (predominant in dilute solution), $FeCl_3(H_2O)_3$ and $FeCl_4^-$ (predominant in concentrated solution).

Studies of the ferric chloride etching of copper, mild (low carbon) steel, spring (high carbon) steel and stainless steels have been made in order to determine the effects of etchant temperature and concentration on etch rate and surface finish. The effects on etch rates are shown in table 4.2.

**Table 4.2**  Dependence of rate of etching on $[FeCl_3]$.

| Material | Temperature of $FeCl_3$ $T(^\circ C)$ | Concentration of $FeCl_3$ for maximum rate of etching at $T^\circ C$ | Rate of etch $(\mu m\,min^{-1})$ | Reference |
|---|---|---|---|---|
| Copper | 25–45 | 2.0–2.5M (29–34 $^\circ$Bé) | — | Burrows |
|  | e.g. 45 | 2.5M (34 $^\circ$Bé) | 5.6 (flowed) | et al. (1964) |
| Copper | 25 | 2.5M (34 $^\circ$Bé) | 31 (sprayed) | Bogenschütz et al. (1979) |
| Mild steel | 60–80 | 2.1–2.3M (30–32 $^\circ$Bé) | — | Maynard |
|  | e.g. 70.1 | 2.3M (32 $^\circ$Bé) | 6.6 (stirred) | et al. (1984b) |
| Spring steel | 50 | 2.1M (30 $^\circ$Bé) | 32 (sprayed) | Hamidon Musa (1984) |
| Stainless steels: AISI 304 (19% Cr, 10% Ni) | 50 | 2.3M (32 $^\circ$Bé) | 38 (sprayed) | Allen et al. (1981) |
| X12 CrNi 17 7 (17% Cr, 7% Ni) | 25 | 2.5M (34 $^\circ$Bé) | 16 (sprayed) | Bogenschütz et al. (1979) |
| X5 CrNi 18 9 (18% Cr, 9% Ni) | 50 | 2.3M (32 $^\circ$Bé) | 25 (sprayed) | Visser et al. (1984) |

It should be noted that at higher etchant concentrations the rate of etching is reduced (§4.3.2) and it has been suggested by Maynard that this phenomenon results from the slower rate of adsorption of $FeCl_4^-$ on the metal such that surface-limited kinetics predominate. Measurement of the enthalpy of activation, $\Delta H = 10.9\ kcal\,mol^{-1}$, also supports the hypothesis, as this value is too high for a diffusion-limited reaction.

It should also be remembered that during a chemical etching reaction the basic laws of electrochemistry must be obeyed. Although no external current is supplied to the reaction site, cathodic and anodic sites are present on the metal surface and the current flowing as a result of metal oxidation must be balanced by the current flowing due to reduction of the etchant species (see figure 4.1).

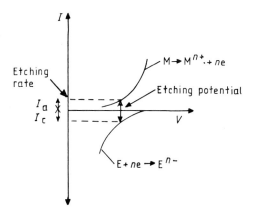

**Figure 4.1** At the etching potential, the rate of oxidation of the metal (M) equals the rate of reduction of the oxidising chemical species (E), the etchant. Note that E may be a cation or anion, e.g. $Fe^{3+}$ or $Fe(CN)_6^{3-}$. $I_a$ is the anodic current, $I_c$ the cathodic current.

As Faraday's laws of electrochemistry apply to the reaction, then the gram equivalent of the metal will be liberated by the passage of one faraday (96 500 coulombs) of charge. This allows the maximum theoretical rate of etching ($R_{max}$) to be calculated from the equation:

$$R_{max} = \frac{iM}{nF \cdot d}$$

where $i$ is the current density (mA cm$^{-2}$), $M$ is the atomic weight of metal (g), $F$ represents one faraday (96 500 A s), $n$ is the valence change and $d$ is the density of metal (g cm$^{-3}$). Therefore,

$$R_{max} = 0.006\,18 \left(\frac{M}{nd}\right) i \qquad (\mu m\ min^{-1}).$$

It can be seen that any increase in current flow will result in faster etching. It should be noted that $n$ is not always easy to evaluate if mechanisms are not known or if the oxidised metal can exist in more than one stable valence state.

If a corrosion diagram is drawn in $V$–$I$ coordinates (figure 4.2) the intersection of the curves will give the corrosion (etching) potential and the corrosion (etching) current.

A fast method of obtaining these two parameters much favoured by corrosion scientists is to allow the metal to attain a steady-state corrosion potential and then apply both positive and negative potentials to the metal to increase anodic and cathodic currents respectively.

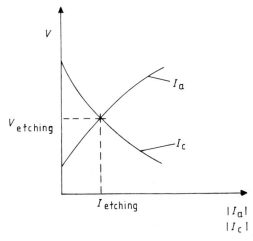

**Figure 4.2** Situation depicted in figure 4.1 redrawn as a corrosion diagram in $V-I$ coordinates.

The etching current ($I_{\text{etching}}$) and etching potential ($V_{\text{etching}}$) may be obtained from Tafel plots of $V$ versus $\log I$ by extrapolation of the two curves to a point of intersection (figure 4.3). The Tafel plot takes the form of a straight line at higher overpotentials as

$$\eta = a + b \log i$$

when

$$-\frac{RT}{F} \gg \eta \gg \frac{RT}{F}$$

and where $\eta$ is the overpotential (volts), $a$ and $b$ are constants, $i$ is the etching current density (mA cm$^{-2}$), $R$ is the gas constant, $T$ is absolute temperature (K) and $F$ is the faraday.

This technique has been used by Gosling (1984, 1985) to measure etching current densities in static and spray-etching of copper with ferric chloride. At 20 °C, results are in excellent agreement with measured weight losses and indicate that spray etching increases the current flow by a factor of 17.

As the rate of etching increases with agitation, it is not surprising, therefore, to discover that all modern etching machines dispense etchant on to the metal substrates by pumping it through banks of spray nozzles to produce turbulence and agitation at the metal surface (§4.6). To maintain constant etching rates, the etchant temperature must be strictly controlled as diffusion controlled reactions are temperature dependent. The diffusion coefficient, $D$, is related to the standard diffusion coefficient, $D_0$, according to the equation

$$D = D_0 e^{-E/RT}$$

where $E$ is the activation energy (usually of the order of 4 or 5 kcal mol$^{-1}$), $R$ is the gas constant (1.98 cal mol$^{-1}$ K$^{-1}$), and $T$ is the absolute temperature (K).

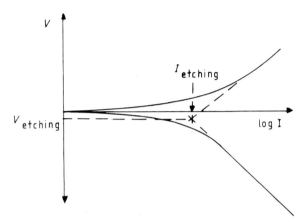

**Figure 4.3** Obtaining values of $I_{\text{etching}}$ and $V_{\text{etching}}$ by extrapolation of the two straight line portions of the anodic and cathodic Tafel plots.

In practice, it is found that for each rise of 10 °C, the etching rate increases by a factor of 1.1 to 1.5. In order to obtain fast etching rates, high temperatures should be used. However, etching temperatures are usually kept below 55 °C, as higher temperatures produce distortions in the PVC plastics materials used in the construction of most etching machines. Highly exothermic etching reactions (see table 4.3) may need cooling and modern production machines contain both heaters and cooling coils.

## 4.3 Ferric chloride

### 4.3.1 Chemistry

Ferric chloride is a red-brown, volatile solid which is very hygroscopic and forms a yellow-brown aqueous solution. The solid compound is usually purchased as the yellow hexahydrate, $FeCl_3 \cdot 6H_2O$.

**Table 4.3** Heat evolution through etching and regeneration processes.

| Metal (M) | (A) Heat evolved on etching† $M_{(s)} + nFeCl_{3(aq)} \rightarrow M^{n+}_{(aq)} + nCl^-_{(aq)} + nFeCl_{2(aq)}$ | | (B) Heat evolved on regeneration† $nFeCl_{2(aq)} + \frac{n}{2}Cl_{2(g)} \rightarrow nFeCl_{3(aq)}$ | | (C) Total heat evolved (i.e. A + B) | |
|---|---|---|---|---|---|---|
| | $-\Delta H$(kcal/gm atomic wt of metal etched) | $-\Delta H$ (Btu/lb of metal etched) | $-\Delta H$(kcal/gm atomic wt of metal etched) | $-\Delta H$ (Btu/lb of metal etched) | $-\Delta H$(kcal/gm atomic wt of metal etched) | $-\Delta H$ (Btu/lb of metal etched) |
| Copper ($n = 2$) | 10.8 | 306 | 53.8 | 1526 | 64.6 | 1832 |
| Nickel ($n = 2$) | 41.5 | 1274 | 53.8 | 1651 | 95.3 | 2925 |
| Iron ($n = 2$) | 47.2 | 1523 | 80.7‡ | 2603 | 127.9 | 4126 |
| Aluminium ($n = 3$) | 164.8 | 11 005 | 80.7 | 5389 | 245.5 | 16 394 |

† These values may be calculated from standard thermodynamic constant tables.

‡ In this reaction three molecules of $FeCl_{3(aq)}$ are regenerated although $n = 2$,

i.e. $Fe_{(s)} + 2FeCl_{3(aq)} \rightarrow Fe^{2+}_{(aq)} + 2Cl^-_{(aq)} + 2FeCl_{2(aq)} \equiv 3FeCl_{2(aq)} \xrightarrow{1\frac{1}{2}Cl_2} 3FeCl_{3(aq)}$.

An aqueous ferric chloride solution is acidic due to hydrolysis of the salt, and it is also a strong oxidising agent, as shown by the value of the standard electrode potential of the half-reaction

$$Fe^{3+} + e^- \rightleftharpoons Fe^{2+} \qquad E_0 = 0.771 \text{ volt.}$$

The chemical reaction involved in the etching of ferrous materials is

$$2Fe^{3+} + Fe \rightarrow 3Fe^{2+}.$$

For copper and its alloys the etching reaction is more complicated. The chemistry involves

| | | |
|---|---|---|
| $Fe^{3+} + Cu \rightarrow Cu^+ + Fe^{2+}$ | at the metal surface | (1) |
| $Fe^{3+} + Cu^+ \rightarrow Cu^{2+} + Fe^{2+}$ | in solution | (2) |
| $Cu^{2+} + Cu \rightarrow 2Cu^+$ | at the metal surface | (3) |

It can be seen that as $[Fe^{3+}]$ decreases, reaction (3) increases in importance. Saubestre (1959) has calculated that at 30% theoretical exhaustion of the etchant, 40% of the copper is etched by cupric ion and 60% by ferric ion, but at 50% theoretical exhaustion, 80% of the copper is etched by cupric ion. However, as $[Fe^{3+}]$ decreases, the overall rate of etching also decreases and, in practice, the etchant is disposed of at low theoretical exhaustion levels or regenerated *in situ*. In a regeneration system (see §4.3.6), $[Fe^{3+}]$ is kept high so that the overall reaction is:

$$2Fe^{3+} + Cu \rightarrow Cu^{2+} + 2Fe^{2+}.$$

It will be noted that due to the highly exothermic reaction of ferric chloride with aluminium (table 4.3) the etchant is best diluted to 12–20 °Bé strength (table 4.1) whereas the etching of ferrous, nickel and copper based materials is carried out with more concentrated solutions.

Ferric chloride cannot be used to etch silver as water-insoluble silver chloride is formed as a reaction product on the metal surface, which consequently prevents further etching. Fortunately silver nitrate is soluble in water and silver may therefore be etched in ferric nitrate solution instead.

### 4.3.2  Effects of dilution

The dilution of $FeCl_3$ etchants can be expressed traditionally, by reference to the Baumé density scale, or, more correctly, by reference to the molarity of the solution.

°Baumé (°Bé) can be calculated by determining the specific gravity (SG) of the liquid with a hydrometer and substituting this value in the formula:

$$°\text{Bé} = 145 \times \frac{(\text{SG} - 1)}{\text{SG}} \qquad \text{(see also Appendix C).}$$

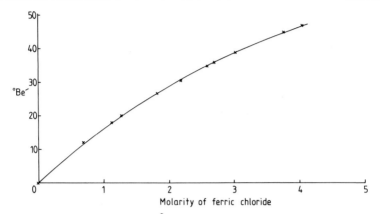

**Figure 4.4** Conversion of °Bé to molar strength of aqueous ferric chloride solutions.

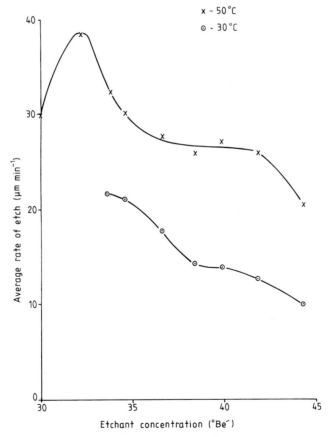

**Figure 4.5** Effect of [FeCl₃] on average rate of etch. Etchant contained 0.20% free HCl. Substrate is AISI 304 stainless steel.

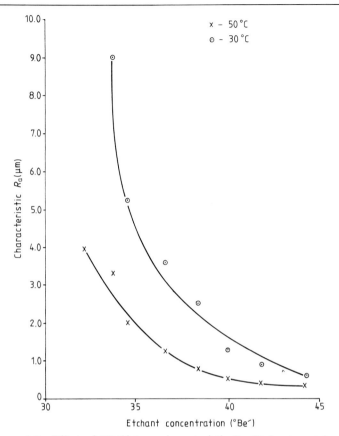

**Figure 4.6** Effect of [FeCl₃] on characteristic $R_a$. Etchant contained 0.27% free HCl. Substrate is AISI 304 stainless steel.

1 M FeCl₃ solution contains 162.2 g l⁻¹ of the anhydrous salt or 270.3 g l⁻¹ of the hexahydrate salt. A graph for converting °Bé to molarity is shown in figure 4.4.

As FeCl₃ etchants are usually purchased in concentrated liquid form, they may need diluting prior to use. This tends to increase the rate of etching (figure 4.5) but will increase the roughness of the etched surface for materials such as stainless steel, as shown in figure 4.6. Maynard has also found a similar trend when etching mild steel in ferric chloride, but the changeover from smooth ($R_a < 0.25\ \mu$m) to rough etching ($R_a > 0.25\ \mu$m) occurs suddenly at critical concentrations, dependent on temperature ($R_a$ is a quantitative measure of surface roughness).

### 4.3.3    Effects of addition of hydrochloric acid (HCl)

Some free HCl is formed by hydrolysis when FeCl₃ is dissolved in water, but a controlled quantity of extra acid is often added to the solution,

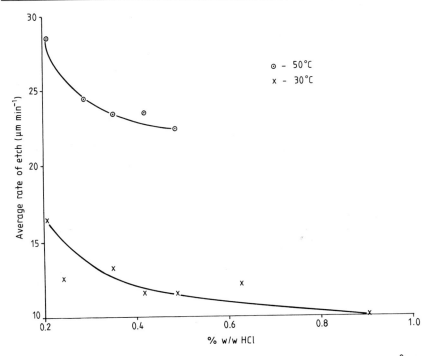

**Figure 4.7**   Effect of [HCl] on average rate of etch. Etchant strength: 40.2 °Bé. Substrate is AISI 304 stainless steel.

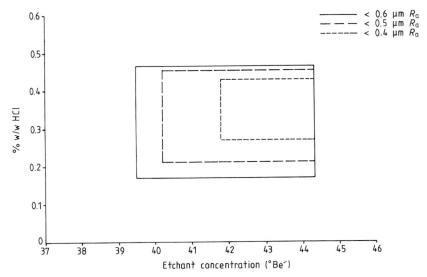

**Figure 4.8**   [FeCl$_3$] and [HCl] selection for obtaining smoothly etched finishes on AISI 304 stainless steel.

especially when etching aluminium, nickel–chromium alloys and steels containing chromium, nickel or cobalt.

Effects of HCl additions appear to be dependent on the substrate metallurgy. For instance, results obtained from etching AISI 304 stainless steel (Allen *et al*. 1981) indicate that HCl additions retard the rate of etching, by forming a diffusion barrier at the metal surface (figure 4.7) and that the amount of free HCl present needs to be carefully controlled if a good surface finish is required (figure 4.8).

However, in etching mild steel, it was found by Maynard that whilst HCl additions also retarded the rate of etching, they had no effect on surface finish!

### 4.3.4   Addition of nitric acid (HNO₃)

Molybdenum can be etched in aqueous $FeCl_3$ but, as the etching rate is very slow, nitric acid based etchants or ferric nitrate are usually preferred. However, spray-etching of steels containing molybdenum can be carried out satisfactorily with aqueous $FeCl_3$–$HNO_3$ mixtures as etchants. Unfortunately, no quantitative data on rates of etching, surface finish and optimum composition appear to have been published.

### 4.3.5   Effects of by-products

If a single etching machine is used for different metals, problems can arise due to some substrates being exposed to certain by-products formed from the etching of dissimilar materials. For instance, stainless steels are particularly sensitive to the presence of copper by-products, the phenomenon revealing itself by producing a poor, rough and pitted surface finish on steel. In practice, therefore, copper is never etched in a machine prior to stainless steel. The flow of work is strictly controlled so that etching of stainless steels is followed by nickel–iron alloys and carbon steels, while copper alloys are only etched at the end of a production cycle. The etchant can then be disposed of, although 'spent' $FeCl_3$ used for etching copper has been used for surface etching and blacking of aluminium (Horne 1974).

As etchant disposal can be very wasteful (see minimum sump capacities in table 4.7) it may be more economic to use a number of etching machines, restricting each to certain types of materials only.

In the USA many companies are now etching ferrous materials in ferric chloride and copper alloys in cupric chloride. Although this means purchasing extra machines, regeneration plants can be fitted to help lower costs.

### 4.3.6  Regeneration of ferric chloride etchant

When etching ferrous materials with $FeCl_3$, regeneration of etchant is possible by treating it with chlorine gas, i.e.

$$2FeCl_3 + Fe \xrightarrow{\text{Etching}} 3FeCl_2$$

$$\downarrow \begin{array}{l}\text{Regeneration}\\[4pt] 1\tfrac{1}{2}Cl_2\end{array}$$

$$3FeCl_3$$

It can be seen from these reactions that 50% extra $FeCl_3$ etchant has been created from iron dissolved from the substrate. This excess etchant can be pumped off from the etching machine sump and stored for use at a later date, or even sold. If, however, ferrous material contains high percentages of alloying elements (e.g. chromium, nickel etc.), regeneration is much less efficient and chlorination can then only be regarded as a rejuvenator.

It is also possible to regenerate with sodium chlorate ($NaClO_3$) and hydrochloric acid solutions, i.e.

$$2FeCl_3 + Fe \xrightarrow{\text{Etching}} 3FeCl_2$$

$$\downarrow \begin{array}{l}\text{Regeneration}\\[4pt] 3HCl + \tfrac{1}{2}NaClO_3\end{array}$$

$$3FeCl_3 + \tfrac{1}{3}NaCl + 1\tfrac{1}{2}H_2O$$

It should be noted that regeneration with chlorine gas can lower costs by a factor of ten (see table 4.4).

**Table 4.4**  Costs of etching 1 lb of iron with 42 °Bé ferric chloride (Wible 1981).

| Regeneration method | US$ (as at March 1981) |
| --- | --- |
| None | 3.00–7.00 |
| Chlorine gas | 0.34–0.81 |
| $NaClO_3$/HCl | 0.74–1.20 |

## 4.4  Cupric chloride

### 4.4.1  Chemistry and usage

As cupric chloride is a rather messy, yellow-brown, hygroscopic solid, it is usually purchased as the green crystalline dihydrate salt, $CuCl_2 \cdot 2H_2O$.

Although cupric chloride etchants can be used in both alkaline and neutral aqueous solutions, the etchant most commonly used in PCM is acidic and comprises the copper salt, water and hydrochloric acid.

The etchant may be used for etching copper, beryllium copper and brass and, *if regeneration is used*, in preference to FeCl₃ as it is more economical (table 4.5).

**Table 4.5**  Costs of etching copper with cupric chloride etchant.

| Regeneration method | Costs US $/lb (Wible 1981) (as at March 1981) | UK £/kg (Decker) |
|---|---|---|
| Chlorine | 0.24–0.58 | 0.19 |
| NaClO₃/HCl | 0.43–0.70 | 0.40 |
| H₂O₂/HCl | 0.77–1.04 | 0.56 |

The cost of etching copper with ferric chloride is $2.56–$5.34 but with regeneration the cost is lowered to $1.00–$2.16. The cost of etching copper with cupric chloride without regeneration is prohibitive.

The standard electrode potential of the half-reaction

$$Cu^{2+} + e^- \rightleftharpoons Cu^+ \qquad E_0 = 0.153 \text{ volt}$$

shows the etchant to be an oxidising agent and the chemical reaction involved in the etching of copper is

$$CuCl_2 + Cu \rightarrow 2CuCl.$$

In an acidic etchant, HCl additions serve to complex the insoluble CuCl and render it soluble, thereby increasing the rate of etching (table 4.6). An additional benefit is that the capacity for dissolved copper is increased.

**Table 4.6**  Effect of hydrochloric acid additions on the rate of etching copper in cupric chloride solution. Pressure = 25 psi (Murski 1981).

| Cupric chloride | Temperature | HCl | Rate of etching ($\mu$m min$^{-1}$) |
|---|---|---|---|
| 33 °Bé | 49 °C | 1N | 25 |
| 33 °Bé | 49 °C | 2N | 31 |
| 33 °Bé | 49 °C | 3N | 38 |

### 4.4.2  Regeneration of acidic cupric chloride etchants

The $CuCl_2$ etchant may be regenerated as follows:

(i) With chlorine gas:

$$2CuCl + Cl_2 \rightarrow 2CuCl_2.$$

This is the most economical method (table 4.5).

(ii) With sodium chlorate and hydrochloric acid:

$$2CuCl + \tfrac{1}{3}NaClO_3 + 2HCl \rightarrow 2CuCl_2 + \tfrac{1}{3}NaCl + H_2O.$$

Although this is cheaper than method (iii) the solution is contaminated with sodium chloride which reduces its salvage value.

(iii) With hydrogen peroxide and hydrochloric acid:

$$2CuCl + H_2O_2 + 2HCl \rightarrow 2CuCl_2 + 2H_2O.$$

This method is used mostly in Europe and although it avoids the use of chlorine gas cylinders, over-chlorination (with subsequent release of free chlorine gas) can result if excessive amounts of $H_2O_2$ and HCl are mixed together.

## 4.5   Other etchants

### 4.5.1   Acidic etchants

Hydrochloric, nitric, sulphuric and hydrofluoric acids can be used as etchants but possess the disadvantage of being much more dangerous than ferric chloride and cupric chloride solutions.

Nitric acid evolves obnoxious gases on reacting with certain metals, but, when diluted, it is the standard etchant for the reactive metals zinc and magnesium and may be used in spray-etching machines:

$$3Zn + 8H^+ + 2NO_3^- \rightarrow 3Zn^{2+} + 2NO\uparrow + 4H_2O.$$

(nitric oxide gas)

Mixing 1 volume of nitric acid with 3 volumes of hydrochloric acid produces the extremely dangerous etchant known as aqua regia, capable of dissolving the noble metals gold and platinum. Due to its hazardous nature, aqua regia can only be used in small volumes with the metals being immersion-etched in a dish or beaker. The chemical reaction is

$$2Au + 3HNO_3 + 9HCl \rightarrow 2AuCl_3 + 3NOCl\uparrow + 6H_2O.$$

(nitrosyl chloride)

The only feasible method of spray-etching gold appears to be with concentrated potassium iodide solutions containing free iodine. However, even this etchant should be handled with care and used with suitable venting.

Molybdenum appears to be particularly susceptible to attack by solutions containing the nitrate anion. Nitric acid mixtures with sulphuric, hydrochloric or hydrofluoric acids have been used for dish-etching of molybdenum but spray-etching systems utilise ferric chloride/nitrate solutions or alkaline etchants (§4.5.2).

The reaction of molybdenum with a nitric–hydrofluoric acid mixture is purported to be (Harris 1976):

$$Mo + 6HF + 6HNO_3 \rightarrow MoF_6 + 6NO_2\uparrow + 6H_2O.$$
$$\text{(nitrogen dioxide)}$$

HF–HNO$_3$ mixtures are also used to etch titanium, as HF solutions on their own can cause hydrogen embrittlement due to the chemical reaction:

$$Ti + 3HF \rightarrow TiF_3 + \tfrac{3}{2}H_2\uparrow.$$

This reaction has been investigated recently by Turner (1985) who estimated 450 calories of heat were generated on etching 1 gram of titanium ($\sim 21.5$ kcal mol$^{-1}$) in 5% HF and suggested that secondary reactions of TiF$_3$ in the etchant are:

$$TiF_3 + 3H_2O \rightarrow \tfrac{1}{2}TiF_3 \cdot TiOF_2 \cdot 5H_2O + \tfrac{1}{2}HF + \tfrac{1}{4}H_2\uparrow$$

$$\downarrow + 3H_2O$$

$$TiO_2 \cdot 4H_2O + \tfrac{5}{2}HF + \tfrac{1}{4}H_2\uparrow.$$

Overall, therefore, the total reaction is

$$Ti + 6H_2O \rightarrow TiO_2 \cdot 4H_2O + 2H_2\uparrow.$$

The reason for adding HNO$_3$ to HF solutions is to oxidise the reaction products and prevent hydrogen gas formation, thus:

$$Ti + 6HF + 4HNO_3 \rightarrow H_2TiF_6 + 4NO_2\uparrow + 4H_2O.$$

Another oxidising agent used for this purpose in the past was chromic acid. However this material presents health hazards (skin dermatitis and ulcerations) and is difficult to dispose of as it contains the pollutant $Cr^{6+}$ ion. Consequently, use of this acid is now much reduced, although commercial, modified solutions are available for etching copper.

It is important to realise that HF is another dangerous etchant that must be handled with great care. Unfortunately, few other materials corrode titanium and tantalum, but Allen and Gillbanks (1985) have used electrolytic etching techniques (§4.7) to successfully fabricate parts from tantalum without fluorides.

### 4.5.2 Alkaline etchants

As aluminium is amphoteric it may be etched not only in acidic solutions but in alkaline solutions also, e.g.

$$2NaOH + 2Al + 2H_2O \rightarrow 2NaAlO_2 + 3H_2\uparrow \qquad -\Delta H = 102.5 \text{ kcal mol}^{-1}.$$

Typically 20% sodium hydroxide is used as an etchant for this strongly exothermic reaction. Unfortunately, evolution of gas can cause uneven etching and in an attempt to overcome this problem modified alkaline potassium ferricyanide solutions have been developed. They had been used previously to etch thin films of chromium and aluminium in microcircuit applications.

In particular, aluminium sheet can be spray-etched in sodium hydroxide–potassium ferricyanide mixtures containing a small quantity of trisodium phosphate to aid chemical polishing of the metal surface (Gerlagh and Baeyens 1975).

Ferricyanide additions suppress the above reaction, which only becomes important at high $[OH^-]$ and gaseous by-products are not formed:

$$Al + 4OH^- + 3Fe(CN)_6^{3-} \rightarrow AlO_2^- + 3Fe(CN)_6^{4-} + 2H_2O.$$

The etchant formulation in table 4.1 has been derived to obtain minimum hydrogen evolution, maximum metal-carrying capacity (only $6 \text{ g l}^{-1}$) and smooth etching, but, in practice, the etch factor (§5.2) achieved is rather low (slightly above 1) and the etch rate is slow even at $55\,^\circ C$ ($9 \ \mu m \ min^{-1}$).

Potassium ferricyanide–sodium hydroxide–sodium oxalate solutions were recommended as etchants for molybdenum as far back as 1966 (Eastman Kodak Co) with in-depth studies of the etchant for use with molybdenum and tungsten being made by Bogenschütz, Braun and Jostan in conjunction with von Beyer in the early 1970s.

The chemical reaction (M = Mo or W) is

$$M + 6Fe(CN)_6^{3-} + 8OH^- \rightarrow MO_4^{2-} + 6Fe(CN)_6^{4-} + 4H_2O$$

and does not involve the sodium oxalate which only acts as a solution buffer to maintain pH.

### 4.6  Etching machines

The etching machines employed with $FeCl_3$ and $CuCl_2$ etchants are usually made of plastics materials such as PVC, with the heaters and cooling coils fabricated from titanium. As PVC distorts at high temperatures, the machines are usually run at a maximum temperature of 55 °C. If temperatures higher than 55 °C are necessary, the machines are constructed from titanium sheet—a very costly item.

Machines may be classified according to whether they process sheets individually (batch etcher) or consecutively (conveyorised etcher); or have static or oscillating spray nozzles.

Etchant is pumped through arrays of nozzles on to the workpiece and to achieve equal rates of etching on both sides of a flat substrate spray pressures can be regulated. If a vertical sheet is etched by two horizontal sprays the pressures should be identical, but in conveyorised machines where the sheet metal is horizontal, the pressure of the lower vertical spray may need to be reduced below that of the upper vertical spray to offset the 'screening effect' of a layer of etchant covering the upper surface of the sheet metal.

After etching, the metal sheets are rinsed with water and dried. A variety of etching machines is listed in table 4.7. Examples of batch and conveyorised etchers are illustrated in figures 4.9 and 4.10.

**Figure 4.9** CRH 50/50 batch etching machine. (Courtesy of CRH Electronics, Solingen, West Germany.)

**Table 4.7** Etching machines.

**(A) Batch etchers**

| Etching machine model (Country of origin) | Maximum size of workpiece (mm$^2$) | Minimum sump capacity (litres) | Number of nozzles per machine | Spray nozzle details | Sheet metal orientation | Comments |
|---|---|---|---|---|---|---|
| Euroelectron E-134 (Holland) | 310 × 330 | 25 | 4 | Horizontal and oscillating | Vertical and static | |
| DEA 30 (USA) | 330 × 355 | 19 | 4 | Horizontal and oscillating | Vertical and static | |
| DEA 30R (USA) | 255 × 255 | 19 | 4 | Horizontal and oscillating | Vertical and rotating | |
| Euroelectron E-2424R (Holland) | 240 × 240 | 25 | 4 | Horizontal and oscillating | Vertical and rotating | |
| CRH 30/30 (West Germany) | 300 × 300 | 250 | 28 | Vertical and oscillating | Horizontal (see comment) | With facility for intermittent rotation of workpiece, max. size of workpiece reduced to 180 × 180 mm$^2$ |
| CRH 50/50 (West Germany) | 500 × 500 | 475 | 85 | Vertical and oscillating | Horizontal (see comments) | With facility for intermittent rotation of workpiece, max. size of workpiece reduced to 300 × 300 mm$^2$ |
| Colight 1010CVR (USA) | 254 × 356 | 27 | 8 | Horizontal and fixed | Vertical (see comments) | With eccentric rotation of workpiece, max. size of workpiece reduced to 254 × 254 mm$^2$ |

100

| | | | | | (see comments) | |
|---|---|---|---|---|---|---|
| Colight 1818CVR (USA) | 635 × 635 | 92 | 24 | Horizontal and fixed | Vertical | With eccentric rotation of workpiece, max. size of workpiece reduced to 457 × 457 mm$^2$ |

**(B) Conveyorised etchers**

| | | | | | | |
|---|---|---|---|---|---|---|
| 24/20E Finishing Services (UK) 30/25E | 610 wide | 164 | 90 | Vertical and static | Horizontal | |
| | 760 wide | 259 | 90 | Vertical and static | Horizontal | |
| 2402 DEA (USA) 2403 | 610 wide | 284 | N/A | Vertical and static | Horizontal | |
| | 610 wide | 379 | | Vertical and static | Horizontal | |
| 503-2 Chemcut (USA, West Germany) 503-3 | 510 wide | 321 | 60 | Vertical and oscillating | Horizontal | |
| | 760 wide | 435 | 112 | Vertical and oscillating | Horizontal | |
| 508-4 Chemcut (USA, West Germany) 547-15 | 1016 wide | 549 | 144 | Vertical and oscillating | Horizontal | |
| | 381 wide | 121 | 32 | Vertical and oscillating | Horizontal | |
| Resco 2-module Bietch system (Italy) | 600 wide | 150 (per module) | 60 (per module) | Vertical fixed followed by vertical oscillating | Horizontal | |
| CRH AES120 (West Germany) | 1000 wide | 300 | 96 | Vertical and oscillating | Horizontal | Effective etch chamber length = 1.0 m |

Comments (rightmost column):
- 2402: Etch chamber 1.29 m long
- (USA) 2403: Etch chamber 1.96 m long

**Figure 4.10**  An eighty-foot long Chemcut etching line (the longest in the world) installed at Koltron Corporation, Sunnyvale, for the PCM of lead frames. (Courtesy of Chemcut Corporation, State College, Pennsylvania.)

**Table 4.8**  Electrolytic photoetching systems.

| Metal | Electrolyte | Reference (Kodak Publication no., page and date) |
|---|---|---|
| Aluminium | Commercial alkaline electrocleaner | *P-7* (2nd ed) p. 35, Oct. 1964 |
| Copper alloys | Ammonium chloride solutions saturated with sodium chloride | *P-7* (2nd ed), p. 38, Oct. 1964 |
| Gold | Alkaline cyanide solutions | *P-7* (2nd ed) p. 41, Oct. 1964 |
| Molybdenum | 10–20% (by weight) sodium hydroxide solution which contains a small percentage of sodium oxalate | *P-91*, p. 23, Dec. 1966 |
| Silver | 15% (by volume) conc. nitric acid in water | *P-7* (2nd ed) p. 50, Oct. 1964 |
| Stainless steel | 25% (by volume) conc. hydrochloric acid in water | *P-91*, p. 31, Dec. 1966 |

**Table 4.9** PCM of difficult-to-etch materials.

| Group metals / Refractory and difficult-to-etch | Spray etching at 55 °C 40 °Bé | | | | Corrosive etchants at 20 °C | | Electrolytic etching | | | | |
|---|---|---|---|---|---|---|---|---|---|---|---|
| | $FeCl_3$ | $Fe(NO_3)_3$ | $KI-I_2$ | $K_3Fe(CN)_6$ | $HF-HNO_3$ | $HNO_3-HCl$ (aqua regia) | 10% HCl (DC) | 4M NaOH (DC) | $CrO_3$ (DC) | $CH_3OH-H_2SO_4$ (DC) | $CH_3OH-H_2SO_4$ (AC) |
| IVa { Titanium | X | X | X | X | ✓ | X | X | X | X | ✓ | ✓ |
| IVa { Zirconium | X | X | X | X | ✓ | X | ? | X | X | ✓ | ✓ |
| Va { Niobium (Columbium) | X | X | X | X | ✓ | X | X | X | X | ✓ | X |
| Va { Tantalum | X | X | X | X | ? | X | X | X | X | ✓ | X |
| VIa { Molybdenum | ? | ✓ | X | ✓ | ✓ | ✓ | X | ✓ | ✓ | ✓ | X |
| VIa { Tungsten | X | X | X | ✓ | ✓ | X | X | ✓ | X | ✓ | X |
| *Noble metals* | | | | | | | | | | | |
| VIII Platinum | X | X | X | X | X | ? | X | X | X | X | X |
| Ib Gold | X | – | ✓ | – | – | ✓ | – | – | – | – | – |
| *Amorphous alloys* | | | | | | | | | | | |
| Vitrovac 6025 ($Co_{66}Fe_4Mo_2Si_{16}B_{12}$) | ? | ? | ✓ | X | ✓ | ✓ | ✓ | X | X | ✓ | X |

*Key* ✓ Etched
? Etched but only with difficulty
X Etchant not suitable (smutting, severe hydrogen embrittlement etc.)
– Information not available

### 4.7   Electrolytic etching

This etching technique is usually reserved only for the difficult-to-etch materials such as refractory and noble metals and corrosion resistant amorphous alloys and superalloys. It is usually carried out on a small scale in beakers (figure 4.11) or in modified electroplating equipment.

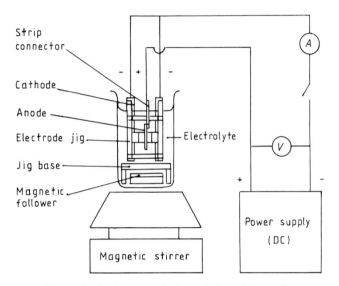

**Figure 4.11**   Layout of electrolytic etching cell.

The metal sheet is connected as the anode in an electrolytic cell containing liquid electrolyte, and the passage of current in the cell effects anodic dissolution.

Table 4.8 shows electrolytic etching systems which have been recommended as alternative methods to chemical etching. The author has been reappraising electrolytic etching in an effort to avoid using toxic chemicals in conventional etching. For instance, parts have been successfully fabricated from Vitrovac 6025, a cobalt-based amorphous alloy, in 10% HCl (Allen and Talib 1984) and from tantalum in 5% $H_2SO_4$ in methanol (Allen and Gillbanks 1985). Some of the electrolytes tested are listed in table 4.9 together with more conventional etchants applied by spraying or immersion.

### 4.8   Bibliography

Allen D M and Gillbanks P J 1985 Manufacture of some SIMS components from tantalum foils by electrolytic photoetching *Precision Engineering* **7** 105–109

Allen D M, Hegarty A J and Horne D F 1981 Surface textures of annealed AISI 304 stainless steel etched by aqueous ferric chloride–hydrochloric acid solution *Trans. Inst. Met. Finish.* **59** 25–9

Allen D M and Talib T N 1984 Electrolytic photoetching of Vitrovac 6025 for the production of magnetic recording heads *Precision Engineering* **6** 125–8

Beyer M, Bogenschütz A F and Jostan J L 1975 Chemisches Ätzverfahren zur Herstellung metallischer Mikrostrukturen aus Molybdän und Wolfram (Part 2) *Metalloberfläche* **29** 506–511

Bogenschütz A F, Braun W and Jostan J L 1975 Chemisches Ätzverfahren zur Herstellung metallischer Mikrostrukturen aus Molybdän und Wolfram (Part 1) *Metalloberfläche* **29** 451–455

Bogenschütz A F, Hostan J L and Mietz H 1979 Verbesserung des Atzfaktors durch Optimierung der Arbeitsparameter beim Atzen von Leiterplatten und Formteilen *Galvanotechnik* **70** 133–143

Burrows W H *et al.* 1964 Kinetics of the copper–ferric chloride reaction and the effects of certain inhibitors *Industrial Engineering and Chemistry, Process Design and Development* **3** 149–159

Decker G Various methods of etchant regeneration—advantages, processes, equipment and cost analysis *Technical Note* (Chemcut GmbH)

Eastman Kodak Co 1964 Kodak Photosensitive Resists for Industry *Publication P-7* (2nd edn) pp. 35, 38, 41, 50

—— 1966 Applications Data for Kodak Photosensitive Resists *Publication P-91* pp. 22, 23, 31

Gerlagh G and Baeyens P 1975 A new etchant for photochemical milling of aluminium *Trans. Inst. Met. Finish.* **53** 133–7

Gosling A D 1984 The investigation of etching mechanistics using a polarization technique *PCMI Journal* no. **18** 9–11

—— 1985 *PCMI Journal* no. **19** 10–12

Hamidon Musa 1984 A study of smut formation in the photochemical machining of aluminium and spring steel *MSc Thesis* Cranfield Institute of Technology

Harris W T 1976 *Chemical Milling—The Technology of Cutting Materials by Etching* (Oxford: Oxford University Press) p. 246

Horne D F 1974 *Dividing, Ruling and Mask-making* (Bristol: Adam Hilger) p. 123

Maynard R B, Moscony J J and Saunders M H 1984a A study of the etching kinetics of low carbon steel using the ferric perchlorate–perchloric acid system as a model *RCA Review* **45** 58–72

—— 1984b Ferric chloride etching of low carbon steels *RCA Review* **45** 73–89

Murski K 1981 Increasing etching productivity *Proc. Tech. Program National Electronic Packaging and Production Conference* 196–207 (Chicago: Cahners Exposition Group)

Saubestre E B 1959 Copper etching in ferric chloride *Industrial and Engineering Chemistry* **51** 288–290

Schlabach T D and Rider D K 1967 *Printed and Integrated Circuitry—Material and Processes* (New York: McGraw-Hill) p. 93

Turner D R 1985 Titanium etching in hydrofluoric acid solutions *J. Electrochem. Soc.* (submitted).

Visser A, Weissinger D and Ullman E 1984 Werkstoffbearbeitung durch Sprühätzen *Galvanotechnik* **75** 14–19; **75** 414–424

Wible P M 1981 Regeneration of etchants *PCMI Journal* no. **6** 7–11

# Chapter 5
# Isotropic Etching of Metals through Photoresist Stencils

# Chapter 5
# Isotropic Etching of Metals through Photoresist Stencils

## 5.1 Isotropic etching

'Isotropic' means 'acting equally in all directions'. Isotropic etching is encountered in nearly all photoetching using liquid etchants and is responsible for producing the characteristic edge profiles of parts made by PCM.

When etching a metal, dissolution of the surface occurs which results in the formation of a sidewall at the photoresist stencil edge (figure 5.1). Once this sidewall has been made there is nothing to prevent the etchant dissolving metal from under the stencil to form what is known as 'undercut' (figure 5.1).

## 5.2 Etch factor and how to measure it

As a quantitative description of the shape of the etched recess, the 'etch factor' can be evaluated from the formula (see also figure 5.2):

$$\text{etch factor} = \frac{\text{depth of etch } (D)}{\text{undercut } (U)}.$$

Etch factor is not a characteristic constant for a given material and etchant formulation but depends on time of etch, original line width in the resist stencil and method of etchant application (§5.3). To be able to etch small holes in metallic substrates it is therefore desirable that undercut should be minimised and photoetching systems giving large etch factors are required.

From figure 5.2 it can be seen that evaluation of etch factor requires the measurement of three quantities, usually the width of the line (or slot) in

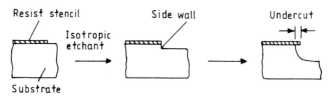

**Figure 5.1**   Etching of a metal surface with an isotropic etchant.

the developed photoresist stencil ($A$), the width of the slot at the metal surface after etching ($B$), and the depth of etch measured when etching is stopped before breakthrough of the metal sheet ($D$). With these measured values, the undercut at each side of the slot is one half of the increase of width produced by etching.

    In practice, measurements of these quantities can be made in different ways, but may be complicated by 'noise' when etching is not ideally uniform. A conceptually attractive method of measuring etch factor is to measure these quantities on cross sections (§6.2) but unfortunately they do not readily show the width of the slot in the photoresist and each section only shows the situation at a single point along a line. Attempts to obtain an average by making many sections from one sample would be excessively laborious. Furthermore, cross sectioning is a destructive method and although sections are indispensable for showing true etch profiles etch factor would be better measured directly on the intact etched metal.

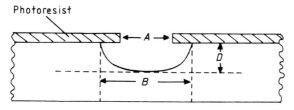

**Figure 5.2**   Etch factor $= D/\frac{1}{2}(B - A)$. Undercut $(U) = \frac{1}{2}(B - A)$.

### 5.2.1   Measurement of required parameters

*Line width of stencil*

After contact printing in a vacuum frame, the width of the line in which the metal is exposed through the photoresist stencil should closely match the line width of the mask. However, the line width of the mask is only an indirect indication of the line width in the stencil as the latter may be affected by light spread or by chemical and physical action on the photoresist during development. For this reason it is better to measure the line width directly on the developed stencil. This may be done either before etching or on a developed stencil on a second piece of metal which is never etched but kept specifically for this purpose.

With some photoresists, the flaps of photoresist stencil (left unsupported by undercutting) retain their shape so well that a good measure of stencil line width can be obtained after etching (figure 5.3). In some instances unsupported photoresist flaps remain even after the metal has been etched right through.

**Figure 5.3** Photomicrograph ( × 62) of metal etched and leaving unsupported flaps of photoresist. The original width of the slot in the resist, width after etching and roughness of the etched edge are shown.

*Etched slot width at the metal surface*

With the vertical illumination in a metallographic microscope and the transparency of a thin layer of photoresist, the position of the etched edge of the metal at the surface can easily be seen without first removing the stencil (figure 5.3). Measurement with an eyepiece scale in this condition has the advantage that disturbing factors such as roughness of the etched edge or local breakdown of the stencil can be readily seen and allowed for. The width at many points along the line can also be determined.

To obtain reasonable accuracy by use of an eyepiece scale, the microscope objectives should be chosen to meet certain conditions. For example, if it is assumed that the position of the edge can be estimated to one fifth of a scale division, the narrowest line that can be measured to 1% must subtend at least 20 divisions and the widest line will subtend 100 divisions of the scale. On these assumptions table 5.1 can be constructed from the scale calibration for the different objectives used ($\times 40$, $\times 10$ and $\times 4$). There is a useful overlap in the line widths measurable with each objective.

**Table 5.1** Scale calibration and measurement errors.

| Nominal overall magnification | Scale division ($\mu$m) | Line width measurable ($\mu$m) | Error ($\mu$m) | Error (%) |
|---|---|---|---|---|
| $\times 40$ | 2.4 | 48–240 | 0.5 | 1–0.2 |
| $\times 10$ | 9.3 | 186–930 | 1.9 | 1–0.2 |
| $\times 4$ | 22 | 440–2200 | 4.4 | 1–0.2 |

*Direct measurement of undercut*

When the undercut flap of resist remains flat and intact (as in figure 5.3) undercut can be measured directly rather than calculated from the difference between the etched width and the line width in the stencil. In this situation, high magnification can be used and very reliable evaluations of undercut should be possible.

*Depth of etch*

In the estimation of depth of etch from cross sections, the original position of the metal surface must be judged by placing the sample so that a line of the eyepiece scale is superimposed on the metal surface at either side of the etched slot (figure 5.4). For this reason the length of the longest lines in an eyepiece scale must correspond to a distance greater than the width of a line. This depends on the design of the eyepiece scale and the magnification of the objective. Assuming that the length of a scale line equals 100 scale divisions, table 5.2 can be constructed.

Acceptably accurate measurements of depth of etch are therefore only obtainable from observations on narrow lines and with deep etching. The narrow lines allow a high magnification to be used, with its attendant small error in estimation of the scale position.

A microscope can also be used to estimate depth of etch by the difference of focus position with the objective focused on the metal surface and on the bottom of the etch pit. In this method, the error is controlled by the depth of focus which is determined by the numerical aperture of the objective.

Although the depth of focus affects the precision of setting both on the metal surface and at the bottom of the etched slot, in table 5.3 it is assumed

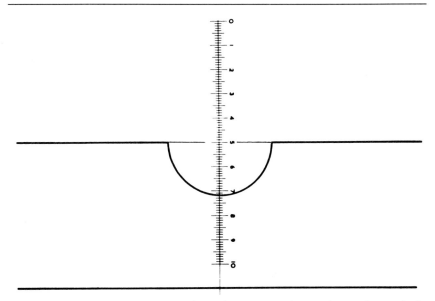

**Figure 5.4**   Diagram showing eyepiece micrometer scale superimposed on etched metal section.

**Table 5.2**   Estimation of depth of etch from cross section.

| Nominal magnification of objective | Scale division ($\mu$m) | Maximum line width ($\mu$m) | Error ($\mu$m) | Depth of etch ($\mu$m) | Error (%) |
|---|---|---|---|---|---|
| × 40 | 2.4 | 240 | 0.5 | 10–100 | 5–0.5 |
| × 10 | 9.3 | 930 | 1.9 | 10–100 | 19–1.9 |
| × 4 | 22 | 2200 | 4.4 | 10–100 | 44–4.4 |

**Table 5.3**   Depths of focus of objectives.

| Nominal magnification | Numerical aperture | Depth of focus (550 nm, in air) | Proportional increase of error |
|---|---|---|---|
| × 40 | 0.65 | 1.2 $\mu$m | × 2.4 |
| × 10 | 0.25 | 8.7 $\mu$m | × 4.5 |
| × 4 | 0.12 | 38 $\mu$m | × 9 |

that the probable error equals the depth of focus. The last column shows the factor by which this error exceeds the estimates of table 5.2.

In contrast to these difficulties associated with optical methods, more promising results can be obtained with measurements made mechanically on a surface roughness measuring machine such as a Taylor–Hobson Talysurf 4. This instrument is designed for examining the roughness of machined metal surfaces. It employs a gearbox to drive a pick-up, fitted with a sharply pointed diamond stylus having a tip width of 2.5 $\mu$m, slowly across the surface. The vertical movements of the stylus modulate a carrier waveform and, after electronic processing, the resultant signal is displayed on a chart recorder. When operated at its lowest sensitivity, this instrument indicates a full-scale deflection of 100 $\mu$m which is a very convenient sensitivity for this study. During initial tests calibration with a slip gauge showed that 100 $\mu$m was measured to within 1 $\mu$m and the trace could be easily read to within 1 $\mu$m.

On etched stainless steel samples Talysurf traces can be made at right angles to the etched slots. The method is non-destructive in that only a faint scratch left by the skid can be seen even on a polished surface. It is simple to make traces at several places along the length of a slot, and the required information is conveniently stored on a chart roll (figure 5.5).

When using a Talysurf 4 it should be noted that due to the geometrical shape of the stylus there is a limit to the depth that can be measured in a narrow line width. However, if the depth of etch does not exceed one half of the line width at the metal surface, it can be shown that a true depth will be recorded by the instrument.

The errors involved in the measurement of depth of etch by the Talysurf method are independent of line width and dependent only on the depth of etch being measured.

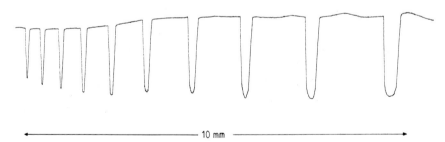

**Figure 5.5**   Read-out from Talysurf 4.

### 5.2.2   Practical evaluation of etch factor on stainless steel

How these procedures work out in practice was tested by Allen *et al.* in a statistical experiment on AISI 302 hard-rolled stainless steel spray-etched for $1\frac{1}{2}$ min in 40 °Bé ferric chloride solution. As seen in figure 5.3 this metal did not give a particularly smooth etch, so the experiment constituted a test of the method of determining etch factor in rather difficult conditions. The

mask used had twelve identical lines 158 $\mu$m wide, and measurements of the
depth of etch and undercut were made at four positions along each line,
giving 48 measurements of each parameter.

In the first treatment of these data, in accordance with the strict definition
of etch factor, the etch factor at each point was determined from the
appropriate measurements of etch depth and undercut. The final value was
the mean of all the individual determinations of etch factor. In the second
method of treatment, a *bar etch factor* was determined from average values
of etch depth and undercut.

In table 5.4 are shown the results of both methods. To determine the
standard deviation for the bar etch factor, standard deviations for both
undercut and depth of etch were determined. Then if mean depth of
etch $= \bar{D}$, standard deviation $= \sigma_{\bar{D}}$ and mean undercut $= \bar{U}$, standard
deviation $= \sigma_{\bar{U}}$, then $\overline{EF}$ (bar etch factor) is defined from the formula

$$\overline{EF} = \bar{D}/\bar{U}.$$

This ratio $\bar{D}/\bar{U}$ has a standard deviation

$$\sigma_{EF} = \frac{\bar{D}}{\bar{U}} \left[ \frac{\sigma_{\bar{D}}^2}{\bar{D}^2} + \frac{\sigma_{\bar{U}}^2}{\bar{U}^2} \right]^{1/2}.$$

The results showed that both methods gave very similar etch factors, but
that the bar etch factor had a somewhat greater standard deviation. This
is not surprising since with the first method it was reasonable to suppose
that variations in depth of etch should be accompanied by related variations
in undercut.

**Table 5.4**   Measurements and standard deviations.

| Method | Mean depth of etch | Mean undercut | Etch factor | Standard deviation |
|---|---|---|---|---|
| Direct calculation and average | – | – | 2.05 | 0.22 |
| Depth of etch by Talysurf 4 | 35.79 $\mu$m | – | – | 3.48 $\mu$m |
| Undercut | – | 17.55 $\mu$m | – | 1.40 $\mu$m |
| Bar etch factor | – | – | 2.04 | 0.26 |

A major difference between the two methods is the procedure required
for collecting the data. For determination of the bar etch factor it is only
necessary to make a number of observations of each parameter at random

positions. For the first method, however, each determination of undercut has to be made at the exact position at which the depth of etch is measured. Fortuitously, this was facilitated by the burnished line produced by the Talysurf skid. If this had not happened, it would have been necessary to mark the samples first and make Talysurf scans and line width measurements at the scribed lines.

## 5.3   Quantitative description of photoetching

The results of photoetching hard-rolled AISI 302 (austenitic 18/8) stainless steel and hard tempered mild steel are presented here as being typical of materials used in PCM. The results form part of a series of papers by Allen, Horne and Stevens under the collective title of 'Quantitative examination of photofabricated profiles' (see §5.6). The data were obtained by making a phototool from specially designed artwork (Stevens) containing parallel lines varying in width from 0.88 to 0.05 mm, imaging into negative-working photoresists dip-coated on to the metal sheets and spray etching through the resultant stencils formed with 40 °Bé ferric chloride at 52 °C. The stainless steel (0.25 mm thick) was cross sectioned to measure the depth of etch $D$, but a Talysurf 4 was used to measure the mean depth of etch $\bar{D}$ of the 0.10 mm thick mild steel as described in §5.2.1.

### 5.3.1   Influence of etch time and line width on the depth of etch

Figures 5.6 and 5.7 show the strong dependence of the rate of downwards etching into the metal on the original line width in the resist stencil. This has been attributed to the difficulty of replacing a spent etchant with fresh solution at the bottom of narrow slots. Without replenishment of the etchant, downwards etching will be slowed down and eventually stop.

The effect of trying to etch narrow and wide features simultaneously in a substrate is illustrated in figure 5.8. For each etch time there is a critical line width in the resist stencil above which the rate of downwards etching is the same. At line widths smaller than the critical line width the downwards rate of etching is retarded. It is apparent therefore that if etching proceeds through stencil apertures with line widths above and below the critical line width then the etched profiles will be different as metal breakthrough will occur at different times. This is discussed in more detail in §5.5.

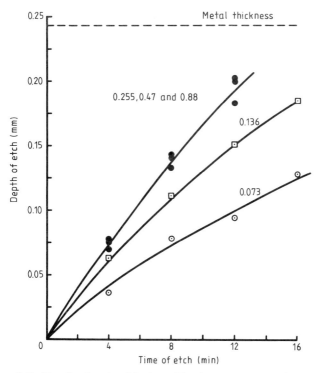

**Figure 5.6** Depth of etch with time. Numbers on curves show width of line (mm) in photoresist stencil. (Stainless steel.)

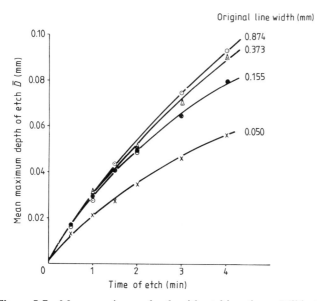

**Figure 5.7** Mean maximum depth with etching time. (Mild steel.)

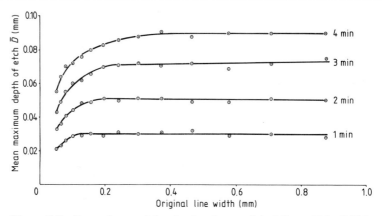

**Figure 5.8**  Dependence of depth of etch on original line width. (Mild steel.)

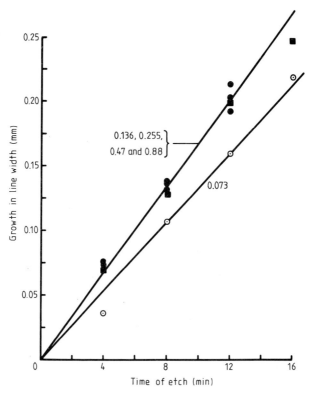

**Figure 5.9**  Growth of line width at metal surface. Undercut (at one edge) equals *half* the growth. (Stainless steel.)

### 5.3.2 Influence of etch time and line width on lateral etching

Figures 5.9 and 5.10 show the results of lateral etching and the formation of undercut with respect to etching time. The growth of line width (calculated as $B - A$ in figure 5.2) is twice the value of the undercut. It can be seen by comparing these results with those displayed in figures 5.6 and 5.7 that the rate of lateral etching is much less dependent on the resist stencil line width than the depth of etch. This is not surprising as lateral etching proceeds fastest along the metal–photoresist interface where spent etchant is rapidly replaced by fresh etchant due to the more favourable hydro-dynamics of the displacement mechanism compared with that required for displacing spent etchant in the bottom of etched recesses (§5.4.1).

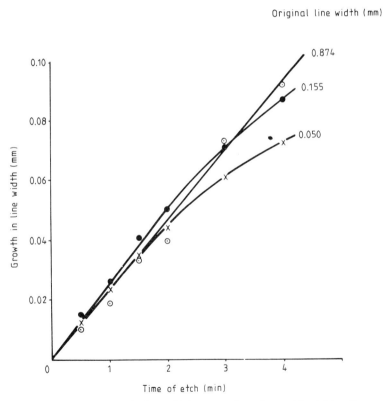

**Figure 5.10**  Growth in line width at metal surface with etching time. (Mild steel.)

### 5.3.3  Influence of depth of etch and line width on etch factor

A plot of depth of etch versus line width is shown in figure 5.11. This type
of result may also be plotted as shown in figure 5.12 where the various
curves emanate from a common origin, but even more information is given

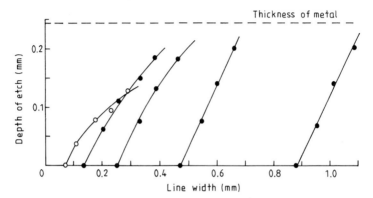

**Figure 5.11**  Interdependence of depth of etch and etched line width.
Resist A and etchant A. (Stainless steel.) See also §5.3.4.

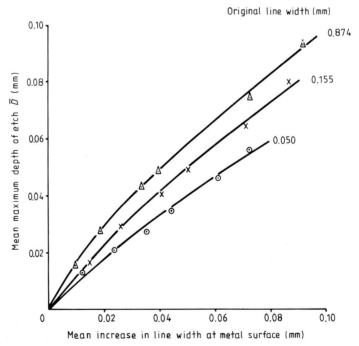

**Figure 5.12**  Interdependence of depth of etch and increase in line
width. (Mild steel.)

by replotting these results as a graph of etch factor (or bar etch factor) versus depth of etch as shown in figure 5.13. Analysis of the resulting curves appears to suggest:

(i) Undercutting cannot start until etching has produced a sidewall (i.e. at very small depths of etch, undercut = 0 so etch factor = ∞).

(ii) As etching progresses, etch factor decreases rapidly with increasing depth of etch and tends to a constant value dependent on the original line width in the resist stencil.

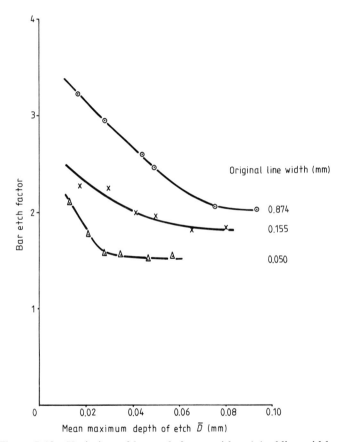

**Figure 5.13** Variation of bar etch factor with *original* line width and depth of etch. (Mild steel.)

It is greatest for the widest lines and least for the narrowest lines. This phenomenon may be directly attributed to the fact that downwards etching into the metal is retarded most in narrow grooves and slots where spent etchant cannot be replenished easily due to restricted access of fresh etchant (§5.4.1).

### 5.3.4 Influence of etchant formulation on etch factor

Figure 5.14 shows the results obtained in attempting to photoetch stainless steel with a poorly formulated etchant (B). This figure should be compared with figure 5.11 where a suitable etchant (A) was used under identical conditions at the same temperature, 52 °C. Considerable undercutting occurs with the poor etchant, B, and furthermore microscopic observation showed excessive undercut had contributed to fracturing of the resist stencil edges. A stronger stencil helps to prevent edge fracturing (see figure 5.15) but will still produce excessive undercut and a correspondingly low etch factor.

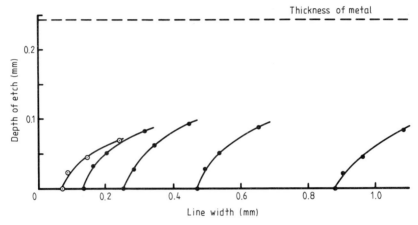

**Figure 5.14** As figure 5.11 but resist A and etchant B.

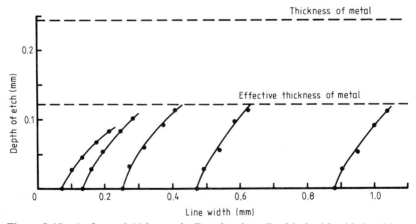

**Figure 5.15** As figure 5.11 but resist B and etchant B with double-sided etching, thus effective metal thickness is halved.

The cause of this problem lies in the fact that etchant A is 40 °Bé FeCl₃ with 0.3% w/w free HCl and etchant B is 40 °Bé FeCl₃ with 0.7% w/w free HCl. However, Allen *et al.* have shown that free HCl concentrations greater than 0.35% w/w will produce a precipitate on the stainless steel surface at 52 °C. This precipitate acts as an etchant barrier and slows down the required etching with a subsequent reduction in etch factor. Confirmation of the importance of a fast rate of etching has been provided by Bogenschütz *et al.* (1979). Experiments suggest that the faster the metal removal rate, the larger the etch factor obtained. It is for this reason also that spray etching is preferred to immersion etching (§4.2).

## 5.4 Theoretical aspects

The data in §5.3 have been derived by practical experimentation with little thought being given to an explanation of the phenomena recorded. In the past few years, however, this situation has been remedied to some extent by the studies of Kuiken *et al.*, in particular on hydrodynamics of etching (§5.4.1) and the development (with time) of etched cavity boundaries (§5.4.2).

### 5.4.1 Hydrodynamics of etching

Pure diffusional transport of an etchant species to a metal surface only occurs in a stationary medium and is responsible for a slow etch rate as the diffusion coefficient is very small ($D \sim 10^{-5}$ cm$^2$ s$^{-1}$; see §4.2). Although etch rate may be increased slightly by raising the temperature (§4.2), spray etching greatly increases this rate by convectional transport of the etchant species.

The sprayed etchant flows over the metal surface and rapidly produces a shallow depression in the surface at apertures in the photoresist stencil. Kuiken suggests that as the depth of the depression increases to form a cavity, the flow of etchant inside the cavity causes one or more vortices (eddies) to be formed as shown in figure 5.16. This causes a marked reduction in etch rate at the bottom of a cavity as the by-products of etching have to diffuse across the streamlines between the vortices to escape from the cavity.

The studies of Alkire *et al.* (1984), based on finite element analysis, reveal that while single eddies are predicted to exist in all but the shallowest of cavities, no secondary eddies are predicted. However, the influence of etchant flow rate on the etchant ion concentration profile near the bottom of the cavity is predicted to decrease as the cavity deepens.

Both of the above theoretical treatments, although slightly different, explain the retardation of downwards (vertical) etching in small holes and grooves as determined experimentally.

Undercutting, which occurs at the metal–photoresist interface, is not retarded as the cavity deepens because this interface is close to the main etchant flow path. As a result it can be seen that etch factor decreases as line widths in photoresist stencils are reduced below a critical width.

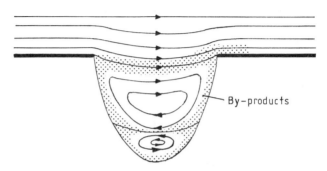

**Figure 5.16**  Etchant flow inside a deep cavity showing vortex formation and flow of by-products. (After Kuiken.)

## 5.4.2  Development of the etched cavity boundary

As etching proceeds with time, the boundary of an etched cavity continually changes as shown in figure 5.17.

A mathematical model of the moving boundary has now been developed by Kuiken (1984) for the very simplest type of etching where the etchant species is transported to the metal solely by diffusion. It has been shown, by using over 130 mathematical equations, that when etching through a

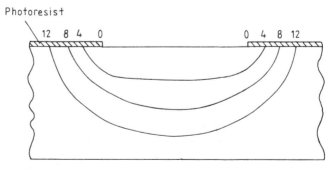

**Figure 5.17**  Development of profile with etching time.

*sufficiently wide* aperture in a resist stencil which maintains rigidity when undercut:

(i) etching occurs faster at the edges of the aperture than in its centre, producing the well-known 'bulge' as shown in figure 5.18;

(ii) the 'bulge' is more pronounced the slower the etching;

(iii) the 'bulge' etch factor ($\sim 1.33$) is independent of the speed at which the moving boundary is displaced in time.

Future work will be aimed at producing a mathematical model describing diffusion etching through narrow resist apertures and the development of computer programs to solve problems of this type (Vuik and Cuvelier 1985).

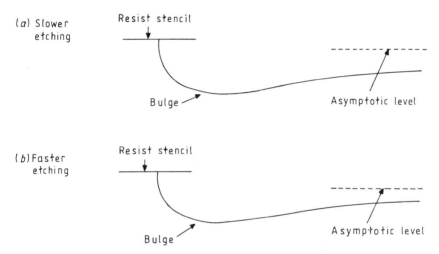

**Figure 5.18** Theoretical profile of a substrate etched through a wide aperture solely by diffusion and showing the 'bulge' phenomenon. The bulge is more pronounced the slower the etch rate. (After Kuiken.)

## 5.5 Implications for component production

We may use the information gathered in §5.3 to produce a better photo-etched product. It is apparent that if a component contains holes and slots of different diameters and widths, etching will produce a different edge profile for each aperture because breakthrough will have been achieved at different times. This undesirable phenomenon may be countered by the use of an etch band (§5.5.1) or some other method of etching where depth of etch is *not* line width dependent (§5.5.2 and §5.5.3). By knowing how hole profiles change with etching times, etched edge quality can also be more easily controlled (§5.5.4).

### 5.5.1  Etch bands

Etch bands are used in PCM to overcome the problem of depth of etch being dependent on line width. By making artwork containing only one specific line width (the etch band width), different sized slots, holes and component perimeters can be etched *at the same rate*. The etch band width is chosen to be equal to or less than the width of the smallest feature. The only problem in choosing small line widths to etch is the resultant low etch factor and slow etching rate. A major advantage of using etch bands is to conserve etchant, as it is only used exactly where it is needed. An example of the use of etch bands is shown in figure 5.19.

**Figure 5.19**   Artwork with etch bands and retaining tabs.

### 5.5.2  Intermittent etching

The research work of Goosen and van Ruler (1976) led to the development of a prototype etching machine in which $FeCl_3$ was sprayed on to copper for 0.4 seconds and then removed again with a blast of air lasting 6 seconds. This intermittent etching cycle results in uniform depths of etch for all line widths in the range 0.015–5.0 mm due to the unhindered access of etchant to the bottom of the etched cavities formed. Although these results look most exciting, unfortunately the technique produces low etch rates ($16\,\mu\mathrm{m\,min^{-1}}$ for copper) due to the cooling effect of the air blast. No production etching machine has yet been made based on the principles of intermittent etching.

The alternative technique of spraying with an atomised etchant was also fraught with practical problems such as immediate evaporation of water from 39 °Bé ferric chloride spray at 50 °C. This was countered by using 29 °Bé etchant at 20 °C, but the technique, though never optimised, proved inferior to intermittent etching.

### 5.5.3  Centrifugal etching

In an attempt to disrupt the etchant flow patterns which are produced in etched cavities (§5.4.1) etching has been carried out in a centrifuge. If the by-products of dissolution increase the specific gravity of the liquid etchant, then a new flow pattern as depicted in figure 5.20 results. This increases etching rate at the bottom of the etched cavity and also reduces undercutting, thereby allowing the etching of holes with diameters less than the thickness of material. Kuiken and Tijburg (1983) have successfully etched circular 0.1 mm diameter holes through 0.2 mm magnesium–bronze using this novel technique.

Etch factors as high as 20 have been recorded when using an ultracentrifuge producing $-25\,000\,g$, and it has been found that etch rate is only weakly dependent on line width (aperture size) in the resist stencil. These characteristics are obviously highly desirable. Unfortunately the disadvantages of centrifugal etching include the non-availability of suitable production equipment and an undesirable widening of holes on the metal surface on completion of etching caused by rapid metal attack at breakthrough.

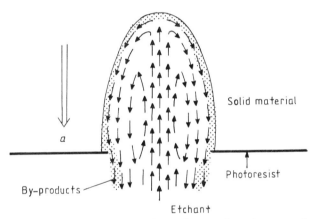

**Figure 5.20** Flow of etchant and reaction products in an acceleration field denoted by $a$. (After Kuiken.)

### 5.5.4 Influence of etch time on edge profiles

The development of edge profiles with etching time may be studied conveniently by cross sectioning (§6.2), taking photomicrographs, projecting the negatives of the photomicrographs on a profile projector, and copying and superimposing the outlines of the profiles on to tracing paper attached to the projector screen.

Profiles of the type shown in figure 5.21 are obtained. When etching from two sides the resultant profiles are 'back-to-back' and (hopefully!) in exact register. After breakthrough of the metal sheet the profile develops as illustrated in figure 5.21. Such profiles ((b), (c) or (d)) are better than that obtained by single-sided etching (figure 5.22) as the deviation from a straight edge profile is less for the same amount of undercut (§6.3).

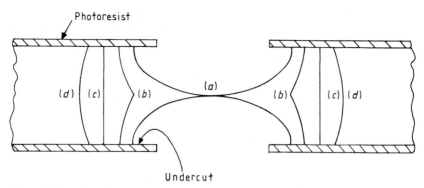

**Figure 5.21**  Development of etched edge profiles. (*a*) Breakthrough point; (*b*) biconvex; (*c*) 'straight'; (*d*) biconcave.

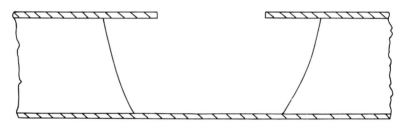

**Figure 5.22**  Single-sided etch profile.

Tapered holes can also be obtained in sheet metal by etching through dissimilar, but registered, stencils. The dissimilarity lies in the fact that wide apertures (slots or holes) are registered with narrow apertures. The practical use made of these etched profiles is discussed in §7.3.1.

## 5.6  Bibliography

Alkire R C, Reiser D B and Sani R L 1984 Effect of fluid flow on removal of dissolution products from small cavities *J. Electrochem. Soc.* **131** 2795–800
Allen D M, Horne D F and Stevens G W W
   'Quantitative Examination of Photofabricated Profiles'
   Part 1: Design of Experiments *J. Photogr. Sci.* **25** 254 (1977)
   Part 2: Photoetched Profiles in Stainless Steel *J. Photogr. Sci.* **26** 72 (1978)
   Part 3: Measurement of Etch Factor *J. Photogr. Sci.* **26** 242 (1978) and reprinted in *PCMI Journal* no. **7** 9 (1982)
   Part 4: Photoetched Profiles in Mild Steel *J. Photogr. Sci.* **27** 181 (1979)
   Part 5: Effect of Stencil Integrity on Etch Factor and the Deep Etching of Stainless Steel *J. Photogr. Sci.* **28** 140 (1980)

Bogenschütz A F, Jostan J L, and Mietz H 1979 Verbesserung des Ätzfactors durch Optimierung der Arbeitsparameter beim Ätzen von Leiterplatten und Formteilen *Galvanotechnik* **70** 133–143

Goosen G and van Ruler J 1976 Intermittent etching: a new possibility for photochemical milling *Interfinish 76: Proc. 9th World Congr. Met. Finish., Amsterdam, October 1976* 17 pp

Kuiken H K 1978 Heat or mass transfer from an open cavity *J. Eng. Math.* **12** 129–155

—— 1984 Etching: a two dimensional mathematical approach *Proc. R. Soc.* A **392** 199–225

Kuiken H K and Tijburg R P 1983 Centrifugal etching: a promising new tool to achieve deep etching results *J. Electrochem. Soc.* **130** 1722–9

Stevens G W W 1977 Sayce Test Chart of Modified Design *J. Photogr. Sci.* **25** 38

Vuik C and Cuvelier C 1985 Numerical solution of an etching problem *J. Comput. Phys.* **59** 247–63

# Chapter 6
# Etching to Dimensional Specifications

# Chapter 6
# Etching to Dimensional Specifications

## 6.1 Introduction

In the etching of metal for decorative applications, cosmetic defects need to be minimised but control of dimensions is usually not critical. However, in the etching of engineering components and devices it is essential that dimensional specifications are rigorously adhered to. The etched components need to be inspected visually for defects and critical dimensions measured accurately. Usually this measurement is achieved with a profile projector and a calibrated scale. The projector allows a relatively large area to be examined but for large single components it may be more convenient to use a universal measuring machine or less expensive micrometers and callipers. A set of titanium plug gauges will be found useful for checking dimensions of circular holes.

The edge profiles obtained by various etching techniques described in §5.5.4 also need to be controlled and although examination under a microscope can be usefully employed to measure deviations from a straight edge profile, the only foolproof method of looking at profiles is by cross sectioning.

## 6.2 Cross sectioning of metal components

A sample preparation technique is needed which meets the following criteria: the component should be sliced across in a plane normal to the plane of the etched foil; the component should be orientated so that the photoetched lines or edges on the surface are normal to the sliced plane; even the most delicate parts of the etched pattern (sometimes constituting isolated wires) should not be distorted by sectioning; and the metal surfaces exposed by sectioning should be worked to a surface flat enough to permit unambiguous location of the metal edge at all points, even for microscopic examination at high magnifications with small depths of focus. Obtaining the required flatness will ensure recording of the true profile.

The accepted procedure for obtaining sections is to embed the sample in a block of synthetic resin, slice across and polish. A high degree of refinement is necessary in cross sectioning thin foils (0.025–1.0 mm thick) as when etched they are rather delicate structures and require embedding (potting) in a medium which does not produce fissures at the resin–specimen interface. To promote resin–metal adhesion, foils need to be cleaned and degreased prior to potting.

Ideally an embedding medium should be a thermosetting casting resin— the elevated temperatures and high pressures needed to use moulding plastics result in distortion of the foil. It should be optically clear and preferably water-white so that the position of slicing can be readily obtained, free from air bubbles which will adhere to the metal and thus weaken support to the foil, free from fissures and cracks in castings 25 mm deep, and hard, as many metals have low abrasion rates which need to be matched to that of the resin.

Keeping the foil vertical may be achieved with two L-shaped supports made by bending a right angle in a rectangular piece of annealed metal, similar in chemical composition to the etched foil (figure 6.1).

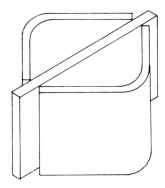

**Figure 6.1** Sample supported by L-pieces, which also aid edge retention.

The embedded part needs to be sliced, ground and polished prior to microscopic examination. Table 6.1 summarises a highly successful procedure used by the author to cross section a variety of stainless steel test pieces and copper alloy components. Even very fine copper mesh, less than 0.006 mm thick, has been successfully examined using 0.3 $\mu$m alpha alumina as a final polishing agent.

**Table 6.1** A successful procedure for cross sectioning metal components.

| Process | Equipment, materials and techniques used |
|---|---|
| (1) Preparation of moulds | Cylindrical, two part, polythene moulds (40 mm diameter and 25 mm deep) were waxed on the cylindrical surface only with Superwax from Trylon Ltd, Wollaston, Northants. |
| (2) Preparation of foils and supporting L shaped pieces | The foils and supports were totally immersed for 10 min in aqueous CD/70 (1 vol + 5 vols water) at room temperature. CD/70 is a metal cleaner and deoxidiser supplied by Chemical Processes Co, Bury Road, Wattisfield, Diss, Norfolk. Cleaning was completed by water rinsing and blow drying with oil-free compressed air. |
| (3) Mixing of the embedding resin | A two-component Araldite® † epoxy system supplied by Plastics Division, Ciba-Geigy (UK) Ltd, Duxford, Cambridge, was used. 30 g of MY753 resin and 5.5 g of HY956 hardener were used per mould and were mixed for 5 min before pouring around the supported foil in the mould. |
| (4) Vacuum degassing | Air bubbles were removed from the resin mixture by evacuating two or three times in a vacuum desiccator. |
| (5) Resin curing | The resin was cured at room temperature for 24 h and the block then ejected from the mould. |
| (6) Slicing | The block was put in the chuck of a cut-off machine with the cylinder axis parallel to the shaft of the abrasive wheel (80 grit carborundum in a resin bond). The block was automatically lubricated during cutting to prevent overheating. |
| (7) Grinding | 220 grit waterproof silicon carbide paper was used for rough grinding in conjunction with a bronze wheel rotating at 500 rpm and water as lubricant. This was followed by successive treatments with 600 and 800 grit silicon carbide papers. |
| (8) Polishing | 6 $\mu$m diamond paste was used on a nylon cloth-covered Bakelite wheel rotating at 250 rpm lubricated with Dialap fluid. All slicing, grinding and polishing materials and equipment were supplied by Metallurgical Services Laboratories Ltd, Betchworth, Surrey. |

† Registered trade mark, Ciba-Geigy.

## 6.3  Edge profile control

In §5.5.4 the influence of etch time on the development of edge profiles has been discussed. Control of profile is therefore achieved by control of etching time. The profile usually requested by engineers is a straight-edge type but etched profiles are usually biconvex, biconcave, convex or irregular taper as illustrated in figure 6.2.

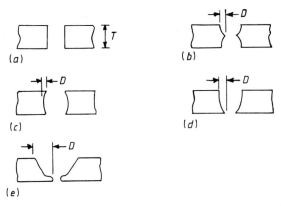

**Figure 6.2**  (*a*) Straight ($D = 0$), (*b*) biconvex, (*c*) biconcave, (*d*) convex and (*e*) irregular taper edge profiles formed in PCM.

The type of profile obtained is dependent not only on the etch time but also on the phototool used for imaging (figure 6.3). The deviation from a straight edge profile ($D$) is dependent on the thickness of the etched metal ($T$) and is least when etching from two sides through registered, mirror-image stencils (figure 6.3 and table 6.2). The deviations listed in table 6.2 are those currently found in commercial PCM and correspond also to those recommended in PCMI Specification D-300.

**Table 6.2**  Relationship of deviation from straight-edge profile ($D$) to etched edge profile in metal thickness $T$.

| Profile | $D_{max}$ (for acceptable profile) |
|---|---|
| Biconvex | $0.2T-0.25T$ |
| Biconcave | $0.2T$ |
| Convex | $0.33T-0.4T$ |
| Taper | variable |

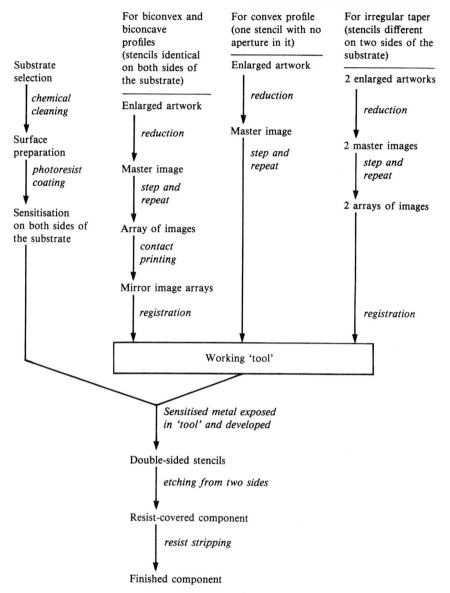

**Figure 6.3** Stages in the production of a photoetched component.

## 6.4 PCM and dimensional control

In order to illustrate how components may be made to specification by control of the PCM process the production of a spring steel shutter blade (figure 6.4) can be considered as typical.

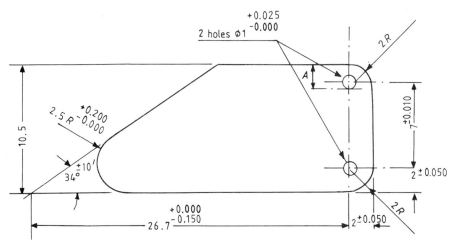

**Figure 6.4**    Spring steel shutter blade (dimensions in mm).

Spring steel is a tough material and when stamped produces burrs (which need to be removed by a lengthy finishing process) and may also deform (figure 6.5(*a*)). Because two pins are inserted into the two 1 mm diameter holes, it is necessary that parts produced by PCM should have reasonably straight edges to the holes.

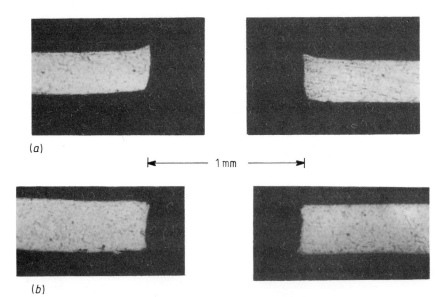

**Figure 6.5**    Cross sections ($\times 186$) of opposite edges of hole in the shutter blade: (*a*) die stamped showing deformation; and (*b*) photo-etched for 150 s.

It is essential, therefore, to find the correct etching time giving an acceptable profile. Artwork corresponding to the part drawing is scribed into 'cut and peel' film on a coordinatograph. This artwork may include an *estimated* etch allowance based on the thickness of material to be etched. In this particular example no etch allowance was made initially.

After manufacture of the phototool the photoresist-coated metal can be imaged, developed and etched for a range of etching times. Microscopic examination or cross sectioning will show the best etching time for the profile desired. In this case the etching time is fixed at 150 s (figure 6.5(*b*)).

### 6.4.1  Change of dimensions with etching time

It will be noticed from examination of the etched blades that hole diameters enlarge as etching time increases and simultaneously the outside dimensions decrease. Some dimensions are etching time independent (e.g. the hole-centre to hole-centre distance, dimension $A$ and the angle of $34° \pm 10'$) and

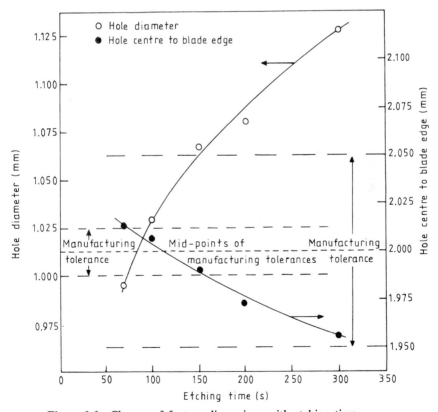

**Figure 6.6**  Change of feature dimensions with etching time.

it will be seen that the rate of change of dimension for a hole-centre to a blade outside edge is approximately one half of that for a rate of change of hole diameter (figure 6.6).

It is now necessary to examine the dimensions with the tightest tolerances that are etching time dependent. In this case they are the hole diameters ($1^{+0.025}_{-0.000}$ mm which is equivalent to $1.0125 \pm 0.0125$ mm), and the hole centre to blade edges ($2 \pm 0.050$ mm). Figure 6.6 shows that at the etching time required to produce an acceptable edge profile (150 s), the hole diameter is oversize by approximately 0.050 mm from the mid-point of the manufacturing tolerance but the hole centre to blade edge dimension is well within tolerance.

It now becomes necessary to produce a new phototool with smaller hole diameters (0.050 mm less). The adjustment is implemented either by redrawing or altering master artwork and producing new phototools after photoreduction and step and repeat printing; or by photoreduction of the original master artwork but with deliberate overexposure to increase line widths, followed by step and repeat printing. This second technique is faster but the amount of 'adjustment' is limited (Stevens 1966).

As the compensation required is 0.050 mm on diameter, if the original master artwork was drawn $\times 20$ oversize then a 1.0 mm reduction on the hole diameter of the master needs to be effected.

### 6.4.2   Etch allowance

The adjustment made to an uncompensated phototool is called an etch allowance. The way in which an etch allowance can be incorporated into the

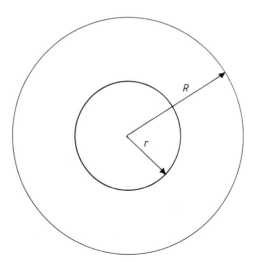

**Figure 6.7**   Washer dimensions.

phototool may be demonstrated by considering the production of a simple washer by PCM (figure 6.7). The dimensions of the two radii, $r$ and $R$, are affected by etching such that a correction to both radii is needed to produce a phototool which can be used to produce washers within specification. The stages involved in production of a modified phototool are shown in table 6.3. Due to the time-consuming nature of these processes etch allowances are usually estimated and incorporated directly into the original master artwork. The estimation is based on the thickness of material to be etched and 'know-how' acquired by familiarity with the material and etchant.

**Table 6.3** Determination of etch allowance and incorporation into a phototool.

| Engineering drawing | Inner radius $r$ | Outer radius $R$ |
|---|---|---|
| Master artwork ($\times 20$ magnification) | $20r$ | $20R$ |
| Photoreduced artwork ($\times 1/20$) and working phototool | $r$ | $R$ |
| Etched washer (outside specification) | $r + c$ | $R - c$ |
| Modified master artwork ($\times 20$) | $20r - 20c$ | $20R + 20c$ |
| Photoreduced modified artwork ($\times 1/20$) and working phototool | $r - c$ | $R + c$ |
| Etched washer (to specification) | $r$ | $R$ |

In the example of the shutter blade, 0.050 mm thick, a 'straight' profile is obtained when undercut is 0.025 mm. In other words the etch allowance $= T/2$ where $T =$ thickness of metal. This relationship may be used as an estimate for most metal/etchant combinations.

### 6.4.3  Estimation of product yield

The reproducibility of the etching process can be determined by measuring identical features etched into the metal for the same duration. Provided the number of measurements is large the standard deviation ($\sigma_p$) for the process can be calculated. For the shutter blade, $\sigma_p$ was found to be 4.5 $\mu$m for the 1 mm diameter holes. With their equivalent tolerance of $\pm 12.5$ $\mu$m (2.78 $\sigma_p$) statistical tables show that a yield of over 99% should be expected purely in terms of correct dimensions.

## 6.5  Bibliography

Allen D M 1981 Design and production of small hole profiles in thin materials *Chartered Mechanical Engineer* (March) 37–40

Allen D M, Horne D F, Lee H G and Stevens G W W 1979 Production of spring steel camera shutter blades by photoetching *Precision Engineering* **1** 25–8

Allen D M, Horne D F and Stevens G W W 1977 Preparation of specimens for microscopic examination of edge profiles *Journal of Microscopy* **111** 203–210

Photo Chemical Machining Institute *What is the photo chemical machining process and what can it do for you?* includes *Publication D-300*: a detailed standard for dimensional tolerances for photo chemical machining, reprinted 1976 (PCMI, 4113 Barberry Drive, Lafayette Hill, PA 19444, USA)

Stevens G W W 1966 Reproduction of aerial images of different sharpness on 'lith type' and extreme resolution emulsions *J. Photogr. Sci.* **14** 153

# Chapter 7
# PCM as a Manufacturing Process

# Chapter 7
# PCM as a Manufacturing Process

## 7.1 Process capability of PCM

### 7.1.1 Limitations of PCM

Many of the limitations to component manufacture by PCM are a direct consequence of the phenomenon discussed in Chapter 5, namely the effect of lateral etching beneath the resist stencil edge. As shown in §§5.5.4 and 6.3, four types of edge profile may be achieved, depending on both etching time and design of the phototooling. However, based on a survey of European contract PCM companies and PCMI Standard Specification D-300, the following limitations have been quantified and may be used as guidelines when assessing the possibility of manufacturing by PCM.

*Minimum hole diameter,* $\varnothing_{min}$

$\varnothing_{min}$ is dependent on metal thickness, $T$. Figure 5.13 shows that least undercut will occur with the smallest depth of etch. In other words, increase in aperture size will be reduced if the thickness of the metal is reduced. Table 7.1 summarises the situation found practically when etching through

**Table 7.1** Relationship of minimum hole ($\varnothing_{min}$) size to metal thickness ($T$).

| Metal thickness $T$ (mm) | $\varnothing_{min}$ |
|---|---|
| < 0.025 | Dependent on conditions |
| 0.025–0.125 | $\geqslant T$ |
| > 0.125 | $1.1T$–$2.0T$ |

145

registered mirror-image stencils. Holes with $\varnothing_{min} \leqslant T$ can only be obtained in metals with $T > 0.125$ mm by etching a taper profile as shown in figure 7.1. Such taper profiles are found in colour TV receiver tube aperture masks (§7.3.1).

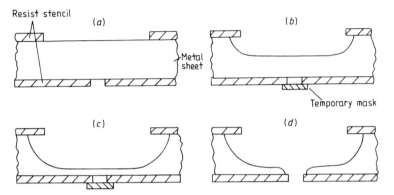

**Figure 7.1** Production of a tapered hole profile showing (*a*) dissimilar registered stencils, (*b*) etching through the larger aperture first, (*c*) substantial thinning of the sheet metal, (*d*) final profile after etching through the smaller aperture. Note that the undercut has been reduced because the metal has been reduced in thickness by the previous etching.

## Minimum metal land width, $W_{min}$

The minimum amount of metal between apertures is also dependent on metal thickness, $T$, as shown in table 7.2.

**Table 7.2**  Relationship of minimum land width ($W_{min}$) to metal thickness ($T$).

| Metal thickness $T$ (mm) | $W_{min}$ |
|---|---|
| < 0.125 | $\geqslant T$ |
| > 0.125 | $\geqslant 1.25T$ |

## Minimum corner radii, $r_{min}$ and $R_{min}$

For an inside corner, $r_{min} = T$ and for an outside corner, $R_{min} = 0.75T$.
    Figure 7.2 shows diagrammatically $\varnothing_{min}$, $W_{min}$, $r_{min}$ and $R_{min}$ and $T$ in a theoretical component.

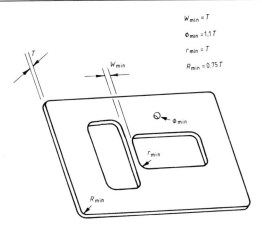

$$W_{min} = T$$
$$\phi_{min} = 1.1\,T$$
$$r_{min} = T$$
$$R_{min} = 0.75\,T$$

**Figure 7.2**  Etched dimensional limitations.

## Maximum material thickness

This is dependent on the metal etched and edge profile requirements. Although the author is aware that 6 mm thick copper and 3 mm stainless steel have been etched successfully, the usual thickness limit for metals is 1.5 mm.

## Dimensional tolerances

In general, a tolerance of $\pm 0.15\,T$ is an accepted estimate of tolerance on etched dimensions, but this will vary according to the metal, sheet size, etchant chemistry and equipment involved.

## Centre-to-centre dimensions

As shown in §6.4.1 hole-centre to hole-centre dimensions are independent of etching time and reflect the accuracy of manufacturing the original artwork. Tolerances are of the order of $\pm 0.05\%$ of the dimension concerned, e.g. $50 \pm 0.025$ mm is typical.

## Bevel, D

The deviations from the classic straight edge profile depend on $T$ and on the phototooling employed. It is desirable to keep $D$ to a minimum for most requirements and for the profiles shown in figure 6.2, $D_{max}$ is listed in table 6.2.

*Range of metals*

A very wide range of metals and alloys can be etched, but not all the metals are etched with equal ease. Ferric chloride and cupric chloride are preferred as etchants by the PCM industry as they are relatively innocuous and can be applied by spraying. Materials which require other etchants (often of a hazardous or toxic nature) are usually immersion etched with a low etch factor as a consequence.

Table 2.1 lists various metals and alloys in approximate order of etchability. Ceramics, glasses, plastics, crystals and semiconductors are usually regarded as being more difficult to etch than metals.

### 7.1.2   Advantages of PCM

Now that the limitations of PCM have been discussed, we can compare the process with other alternative fabrication methods such as photoforming, fine blanking and piercing (also known as high precision stamping) and microdrilling, as shown in table 7.3. Photoforming involves electroplating of metals on to an electrically conductive mandrel which is patterned over its surface with an insulating resist stencil. The photoformed component is removed from the mandrel after plating to the required thickness.

*Economics of PCM compared with stamping*

Manufacture of a part by PCM requires consumables such as cleaners, photoresists, etchant and strippers to be included in the part cost ($P_E$) which is therefore relatively high. Artwork and phototooling also need to be made but this involves relatively low costs ($A$). These economic factors contrast strongly with those of stamping where a high tooling cost for the manufacture of a punch and die ($D$) is associated with a very low part cost ($P_S$). It is apparent, therefore, that PCM is cheaper for low volume production while stamping is cheaper for high volume production. The break-even quantity ($Q$) is dependent on the geometric complexity of the part and is calculated from the formula:

$$Q = \frac{D - A}{P_E - P_S}.$$

Assuming PCM and stamping produce parts to identical specifications, PCM becomes increasingly economic as part complexity increases (see figure 7.3 and table 7.4).

**Table 7.3** Technical comparison of fabrication techniques.

| | Photographic methods | | Machine shop methods | |
| --- | --- | --- | --- | --- |
| | PCM | Photoforming | Stamping | Microdrilling |
| Maximum material thickness | 1.5 mm (6 mm for low resolution work) | 2 mm | 13 mm | Depends on drill size |
| Deviation from a straight profile | < 20% of material thickness ($T$) | Dependent on material thickness ($T$) to resist stencil thickness | A slight taper with a burr | A slight taper with a burr |
| Minimum aperture size | $\varnothing = 1.1T$ for most metals (but not an absolute limit) | $\varnothing = 0.1T$–$0.5T$ | $\varnothing = 0.5T$ (low carbon steel) $\varnothing = 0.75T$ (high carbon steel) | $\varnothing = 0.025$ mm |
| Material | All metals (but vary in etchability) | Usually nickel, copper, silver or gold | Non-brittle metals | Non-brittle metals |
| Process advantages | (1) Produces burr-free and stress-free components (2) Physical and chemical characteristics of metal not altered during processing (3) Variable edge profile | May be the only viable process capable of achieving very high resolution | (1) Forming operations can be carried out whilst blanking (2) Fast | Fast |
| Process disadvantages | (1) A multi-stage process (2) Thickness limitation | (1) A multi-stage process (2) Restricted range of metals (3) Thickness limitation | (1) Long lead times (2) Deburring required | (1) Deburring required (2) Drill fragility (3) Difficult to locate drill (4) Skilled operatives required |

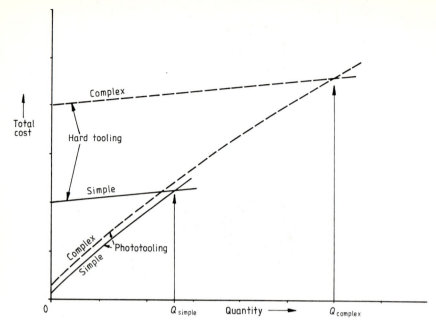

**Figure 7.3**   Typical economics of the production of simple and complex parts by PCM and stamping showing break-even quantity ($Q$) increasing with part complexity.

*Lead times*

PCM lead times are short—days or weeks only, rather than months necessitated by stamping. This big advantage arises because phototooling can be made rapidly whereas stamping requires precision tooling to be made in very tough materials such as tungsten carbide. The manufacture of a punch and die requires a good deal of skilled labour as well as time. As both are costly items, this helps to explain the high price of hard tooling.

*Burrs and tabbing*

One of the advantages of an etched component over a stamped one is that burrs are not present on etched edges, thereby eliminating a need for abrasive treatments to finish the component. If burr-free, ultrasmooth edges are specified, then it is necessary to separate components completely from the parent metal. If tabbing (§1.10) is used, breaking the components out of the sheet will produce small (0.125 mm) 'pips' of metal at component edges. The conventional way of separating components completely from the rest of the sheet is to etch them in a horizontal conveyorised etching

**Table 7.4** The effect of piece-part complexity on the economics of production by PCM and stamping. (Figures by courtesy of Tecan Components Ltd, Weymouth.)

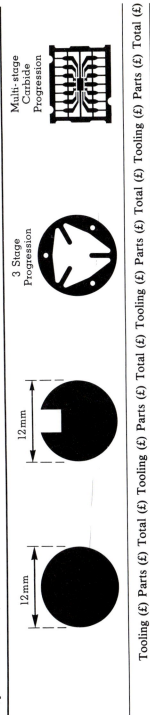

|  | Tooling (£) | Parts (£) | Total (£) | Tooling (£) | Parts (£) | Total (£) | Tooling (£) | Parts (£) | Total (£) | Tooling (£) | Parts (£) | Total (£) |
|---|---|---|---|---|---|---|---|---|---|---|---|---|
| **100 parts** | | | | | | | | | | | | |
| Stamping | 20 | 7 | 27 | 180 | 7 | 187 | 450 | 10 | 460 | 25 000 | 30 | 25 030 |
| PCM | 45 | 20 | 65 | 50 | 20 | 70 | 85 | 20 | 105 | 300 | 55 | 355 |
| **5000 parts** | | | | | | | | | | | | |
| Stamping | 100 | 85 | 185 | 300 | 85 | 385 | 800 | 100 | 900 | 25 000 | 300 | 25 300 |
| PCM | 65 | 90 | 155 | 70 | 90 | 160 | 95 | 95 | 195 | 320 | 600 | 920 |

machine with the metal sheet supported on a plastic mesh to retain the parts.

*Sheet metal properties transferred to components*

PCM does not affect physical or chemical properties inherent to the sheet metal. This is very important where stress, high temperatures or loss of magnetic permeability must be avoided during manufacture. Rival processes such as laser cutting produce heat-affected zones at the cut edges while stamping affects magnetic permeability and can also cause stress. PCM can be regarded as a very gentle process, only affecting the substrate where the etchant chemically reacts with it.

*Three-dimensional etching*

Although PCM is usually associated with the production of planar components, it may also be used to etch the surfaces of cylinders and spheres (§7.3.10). To produce high resolution complex-shaped recesses on these surfaces by other methods is not easy, probably involving engraving and/or milling techniques.

*Design considerations*

The adage 'If you can draw it, PCM can make it' is basically true, provided that the limitations discussed in §7.1.1 are taken into consideration. Certainly the process is ideal for creating complex metal parts, especially those involving arrays of apertures and filigree work.

If designs need to be altered as a result of prototype trials then PCM has obvious advantages as the prototype production method. Modifications are carried out by altering or redrawing artwork and production of new phototooling. This can be done quickly and at little cost, whereas modifications to hard tooling will take longer and will be expensive.

*Prototype to production*

As components in production quantities are made in exactly the same way as prototypes, no variations will be noted in quality after the change-over. Phototools are easily stored, do not wear and repeat orders may be serviced as soon as sensitised metal can be prepared.

## 7.2  Secondary operations

Usually the PCM process finishes when the resist stencil is stripped from the etched metal sheet. However, some resist stencils develop adhesive properties when heated which may be utilised to bond together laminated components such as magnetic recording heads. Stencil-coated arrays of etched and tabbed heads in sheet format, called frets (figure 7.4), are stacked on register pins to ensure exact alignment. The frets are then put in a press and heated under pressure to effect a permanent bond betweeen them. They are then guillotined and the individual recording head blocks carefully ground to the desired size.

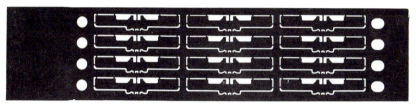

**Figure 7.4** A fret of magnetic recording head laminations. Note the geometry of the register holes.

Lamination bonding is one method of achieving small apertures in a thick component using PCM and, if dissimilar laminations are bonded together, complex three-dimensional channels, pipes and tapered holes can be manufactured *en bloc*.

Other secondary operations, often carried out by contract PCM companies as a service to their customers include: reaming of etched holes for greater precision; drilling of individual holes where the hole diameter is less than that which can be obtained by PCM; forming; surface treatments such as shot blasting; metal finishing such as electroplating, vacuum coating, anodising, chemical blackening (especially useful for optical components), polymer coating, silk-screen printing, painting and filling; or heat treatments to alter the temper of parts.

## 7.3 Functional product range

Functional products made by PCM are found in the electronic, automotive, aerospace, telecommunication, computer, optics, medical, nuclear, metal working and precision engineering industries.

Recognition of the advantages and limitations of PCM, as discussed in §7.1, leads one to conclude that PCM is: the best method for production of prototypes; excellent for small and medium batch production; excellent for volume production of complex parts (e.g. lead frames); the only manufacturing method for components of extreme complexity (e.g. colour TV receiver tube aperture masks); and economically viable for the production of both simple and complex components in difficult-to-stamp materials (e.g. molybdenum). The types of products made by PCM are shown in table 7.5, and some of them will be studied in detail below.

**Table 7.5**  Types of products made by PCM.

| Components containing arrays of perforations | Components with complex geometries | Surface-etched components (§7.3.8) |
|---|---|---|
| grids | levers | rules |
| filters | diaphragms | scales |
| meshes | shims | clutch plates |
| screens | gaskets | emitter contacts (§7.3.9) |
| shaver foils | washers (§7.3.2) | cylindrical and spherical bearings (§7.3.10) |
| colour TV aperture masks (§7.3.1) | springs | edge filters |
| light attenuators | links | hybrid circuit pack lids (§7.3.11) |
| | brackets | boxes and enclosures (fold lines) (§7.3.12) |
| | contacts | nameplates (§7.4) |
| | connectors | decorative plaques (§7.4) |
| | probes | components etched with logos, trademarks, part numbers or instructions |
| | heat ladders, plates and sinks (§7.3.3) | potentiometer leads |
| | lead frames (§7.3.4) rotor and stator laminations | instrument cases |
| | magnetic recording head laminations (§7.3.5) shutter blades iris leaves graticules (§7.3.6) deposition masks | |
| | tape carriers strain gauges combs ⎫⎬⎭ on insulated backing material | |
| | light chopper discs (§7.3.7) encoder discs jewellery (§7.4) | |

### 7.3.1 Colour TV receiver tube aperture mask

Commonly known as a shadow mask, this perforated component is found between the arrays of red, blue and green phosphor stripes or dots on the

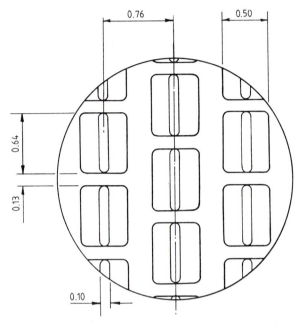

**Figure 7.5** Typical dimensions of a shadow mask (mm).

**Figure 7.6** Cross section of a shadow mask showing tapered hole profile.

inside of the TV screen and the three electron beam sources which activate them.

Figure 7.5 shows typical dimensions of slots etched in 0.15 mm thick mild steel. The mask comprises some 300 000 slots, each of which must be perfect. A cross section through a slot shows it to have a tapered profile as shown in figure 7.6. This tapered profile is produced by etching through registered but dissimilar resist stencils, as shown schematically in figure 7.1. The shadow masks cannot be made by any other method and, as the customer demand is vast, they are made on a continuous 24 hour process.

Further details are found in the articles by Holahan (1965) and the anonymous report in *American Machinist* (1971).

### 7.3.2   Spider washer

This component (figure 7.7) was made in the author's laboratories from various samples of spring steel and was fitted into a prototype industrial device to determine the most suitable thickness and temper of spring steel for the washer based on results of fatigue tests. PCM was the favoured method of production because of the short lead time and low cost.

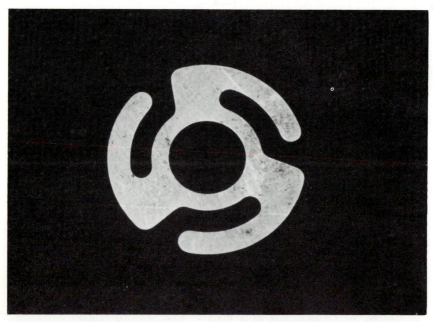

**Figure 7.7**   Spider washer made in several gauges of spring steel for fatigue testing.

### 7.3.3 Heat plate

Figure 7.8 shows a 1.63 mm thick dull nickel-plated copper heat plate. The geometry is complex and many areas are recessed (§7.3.8) to allow integrated circuits with short connector pins to straddle the heat plate and fit into the holes in the underlying printed circuit board.

**Figure 7.8** A 1.63 mm thick, dull nickel-plated copper heat plate. (Courtesy of Tecan Components Ltd, Weymouth.)

### 7.3.4 Integrated circuit lead frame

An integrated circuit (IC) lead frame (figure 7.9) is another example of an electronics component with a complex geometry. High volumes are required but, as IC designs change so rapidly, PCM is favoured as the production method because lead times are short, thus allowing the IC to come on to the market quickly. Some lead frame manufacturers are now etching their products on a line process similar to that devised for fabricating shadow masks (§7.3.1).

### 7.3.5 Magnetic recording head laminations

The laminations are made from magnetic materials such as HyMu 80 and are etched, rather than stamped, because stamping adversely affects the magnetic permeability of such materials.

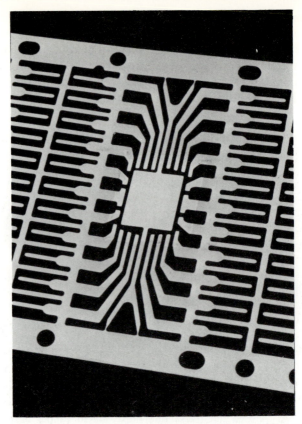

**Figure 7.9**   Part of a strip of spot-plated IC lead frames. (Courtesy of Koltron Corporation, Sunnyvale, California.)

Figure 7.4 shows the format of a narrow fret (§7.2) made in the author's laboratories for the manufacture of a magnetic recording head from a ribbon of amorphous (glassy) metal. This material (Vitrovac 6025) could not be stamped, as it shatters, and could not be etched in ferric chloride either. Electrolytic photoetching, an unconventional etching method in which the substrate is dissolved by connecting it as an anode into an electrolytic cell, was used therefore to etch this novel material (§4.7).

### 7.3.6   Metal graticule

The camera viewfinder graticule shown in figure 7.10 is an interesting component in that it shows the importance of the role of the design engineer in production. The 'wire' frame is deliberately barrel-distorted to counteract pincushion distortion in the cheap viewfinder optics. This simple design solution to the problem was obviously more cost-effective than redesigning the optics, but stamping would have been extremely difficult and costly due to the the shape and narrow width of the frame. PCM provided the economic solution to the problem.

**Figure 7.10**   Plan view of 0.13 mm thick brass graticule ( × 1.5). Note the deliberate 'barrel' distortion of the wire frame. (Courtesy of Kodak Ltd, Harrow.)

### 7.3.7   Light-chopper disc

The example shown in figure 7.11 is made by etching from both sides through mirror-image, registered stencils. This 100 mm diameter, 0.4 mm thick brass disc has 120 slots, 1 mm × 3 mm long, equally spaced around its circumference. Five other apertures are also etched into the disc, the central

**Figure 7.11**   A 100 mm diameter, 0.4 mm thick brass light-chopper disc. (Courtesy of Tecan Components Ltd, Weymouth.)

one having an irregular shape required for attaching the disc to a shaft. Economics favour PCM as a manufacturing method, especially as this component is only required in small quantities.

### 7.3.8 Surface-etched components

Figure 7.12 illustrates how, whilst etching through sheet metal to 'chemically blank' a component, it is also possible to etch the metal surface at the same time by use of registered, but dissimilar, stencils.

Surface etching is used: to mark parts with identification symbols such as alphanumeric characters, logos and trademarks. To enhance contrast,

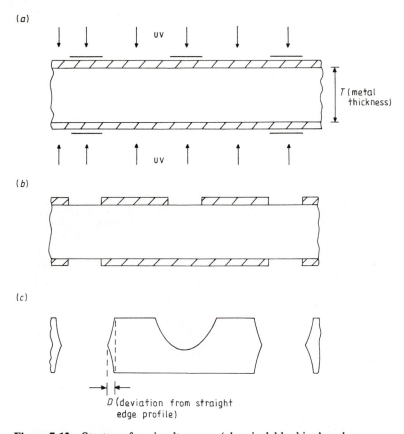

**Figure 7.12** Strategy for simultaneous 'chemical blanking' and surface etching. (a) Negative-working photoresist in contact with a registered phototool. (b) Resist stencil on metal sheet prior to etching. (c) Etched metal component with additional surface etching and after resist stripping.

the etched recesses may be filled with pigmented epoxy or acrylic plastics or various types of paint, or chemically coloured or plated; to produce recessed areas (see §7.3.9 and §7.3.10); to form lipped edges (see §7.3.11); and to form bend or fold lines, so that a third dimension can be incorporated into the component if required (see §7.3.12).

### 7.3.9 Emitter contacts

These parts etched from molybdenum have intricate geometries as shown in the example of figure 7.13. They are used in power bipolar transistors, used in such applications as motor control, switching up to 600 amps each and therefore requiring packaging capable of removing large quantities of heat and preventing damage to the silicon element due to differential thermal expansion at temperatures up to 200 °C.

**Figure 7.13**   Molybdenum emitter contact. (Courtesy of Molypress Ltd, Calne.)

### 7.3.10 Spherical pivot bearing

The pivot bearing shown in figure 7.14 is a superb example of three-dimensional etching and again illustrates the versatility of PCM for solving production problems. Phototooling also needs to be spherical so that photoresist–emulsion contact is maintained. Kodak literature suggests this phototooling can be made from imaged acetate-based film softened in acetone and water (1 : 1 by volume) and vacuum formed around the sphere until the film has regained its initial firmness.

Cylindrical bearings are much easier to etch, the phototooling only requiring to be wrapped tightly around the photoresist-coated cylinder, with exposure effected by rotating the cylinder and phototooling in front of a narrow slit of ultraviolet light.

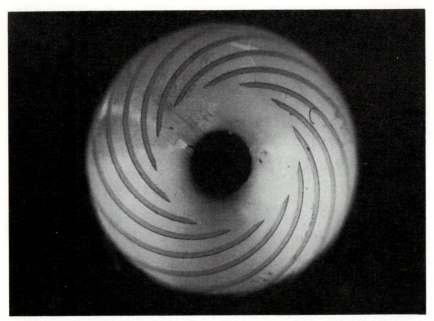

**Figure 7.14**   End-on view of a spherical pivot bearing. (Courtesy of NTN Toyo Bearing Co, Shizuoka, Japan.)

### 7.3.11   Hybrid circuit pack lids

For certain computer and military applications, hybrid circuits need to be packaged into hermetically sealed packs. The packs are made of gold-plated Kovar (an iron–nickel–cobalt alloy with a coefficient of thermal expansion matching that of glass) with pin connectors passing into the pack via sintered glass seals. The pack is made of two parts: the base with the pins and a flanged lid (figure 7.15). After connecting the circuit into the base, the lid is roller seam welded on to the base in an inert atmosphere. The welding process requires that the edges of the lid be smooth, dimensionally accurate and the lip depth extremely uniform—a condition achieved by carefully controlled etching of the costly 0.38 mm thick Kovar sheet.

**Figure 7.15** Gold-plated hybrid circuit pack showing etched lip of the lid. (Courtesy of Tecan Components Ltd, Weymouth.)

### 7.3.12 Boxes and enclosures

Figure 7.16 shows an example of a box fitted with a sliding lid both in its folded form prior to soldering and in the flat sheet as formed after etching and stripping of the resist stencil. Other products utilising fold lines are radio frequency (RF) shields and protective housings.

### 7.4 Decorative and graphic products range

These products are becoming increasingly significant to the PCM industry. Although they are not usually incorporated into component assemblies, with the result that dimensional tolerances are often generous, visual appearance is of the utmost importance. The human eye is extremely critical and consequently great care must be taken to ensure absence of cosmetic

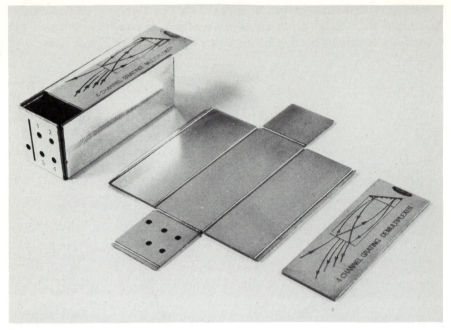

**Figure 7.16**   Folded box and its component parts all made from 1.0 mm thick brass by PCM. (Courtesy of Photofabrication (Services) Ltd, St Neots.)

defects. Attention to surface finish, resolution, colour and contrast is essential.

In the past such products have been made on a labour-intensive basis employing skilled craftsmen and techniques such as engraving. By contrast, PCM allows large numbers of products to be made from one phototool, which can be produced quickly and at low cost. The overall effect of PCM, therefore, is to produce high quality products at relatively low cost. The products include:

### 7.4.1   Jewellery and decorative etchings

Precious metals such as gold and platinum are difficult to etch, but silver and cheap metals such as brass (commonly used in the manufacture of costume jewellery) are easy to etch. PCM is particularly suited to producing complex patterns, shapes and filigree work. These desirable characteristics should be borne in mind by the designer and put to advantage accordingly. Figure 7.17 shows an example of this type of work. Brass may be plated with silver and gold not only to make costume jewellery but also Christmas tree hanging decorations and table centre-pieces (figure 7.18).

**Figure 7.17** Silver jewellery made by PCM and filled with enamel. (Courtesy of Dust Jewellery, Fife.)

**Figure 7.18** Hanging decoration made in gold-plated brass. (Courtesy of Chemart Co, Lincoln, Rhode Island.)

### 7.4.2 Graphic products

Details are etched into the surface of a flat sheet of metal through a resist stencil. To enhance contrast another metal of dissimilar colour may be plated into the etched recess, or it may be chemically coloured or filled with a pigmented medium (§7.3.8).

Products in this category include etched line drawings, maps, plaques, clock faces and sundials as well as notices containing instructions and information. In addition to being aesthetically pleasing, the products are also permanently marked, unlike their silk-screen printed equivalents. If nameplates and notices are to be sited outdoors, they are usually made of corrosion resistant stainless steel or anodised aluminium (figure 7.19).

**Figure 7.19** Promotional calendar in stainless steel. (Courtesy of Hirai Seimitsu Corp, Osaka, Japan.)

### 7.4.3 Models

Intricately detailed models of railway rolling stock, bridges, towers, fences and buildings are examples of typical products successfully made by PCM.

It is of interest to note that in the film *Superman II* the Eiffel Tower scene was obtained by using a model tower predominantly photoetched from brass.

### 7.4.4  Metal sculptures and collages

Due to the ease with which complex shapes can be produced PCM is now being used more by artists working in metals. Visual impact may be enhanced by metal finishing techniques such as chemical colouring and lacquering or other surface treatments.

## 7.5  Bibliography

Allen D M 1981 Design and production of small hole profiles in thin materials *Chartered Mechanical Engineer* (March) 37–40
—— 1984 Photochemical machining—a cost effective rival to metal stamping for parts manufacture *Proc. 1st Int. Conf. Advances in Manufacturing, Singapore, October 1984* pp. 175–187
—— 1985 Solving manufacturing problems by combining photofabrication with surface treatment and finishing techniques *Trans. Inst. Met. Finish.* **63** 160–2
Anon 1971 Chemical milling precision parts *American Machinist* (Feb. 8) pp. 50–52
Eastman Kodak 1968 Chemical Milling with Kodak photosensitive resists *Publication P-131* (New York: Eastman Kodak Co.) p. 20
Holahan J F 1965 Manufacture of colour picture tubes *Electronics World* (December)
Photo Chemical Machining Institute *A photo chemical machining showcase* (PCMI, 4113 Barberry Drive, Lafayette Hill, PA 19444, USA)
—— *What is the photo chemical machining process and what can it do for you?* includes *PCMI D-300*: a detailed standard for dimensional tolerances for photochemical machining, reprinted 1976 (PCMI, 4113 Barberry Drive, Lafayette Hill, PA 19444, USA)
Sato T and Takahashi Y 1980 Photo-chemical milling of embossing rolls *Bull. Jpn Soc. Prec. Eng.* **14** 115–6
Van Delft J Ph, Van der Waals J and Mohan A 1975 Electroforming of perforated products *Trans. Inst. Met. Finish.* **53** 178–183

# Appendix A
# SWG (Standard Wire Gauge) Metal Thickness Conversions to Metric and Imperial Measurements

| Standard wire gauge | Fractions of inch | Decimal equivalents inch | Metric equivalents millimetres | Standard wire gauge | Fractions of inch | Decimal equivalents inch | Metric equivalents millimetres |
|---|---|---|---|---|---|---|---|
| 1 | | 0.300 | 7.620 | 10 | | 0.128 | 3.251 |
| | $\frac{19}{64}$ | 0.2968 | 7.538 | | $\frac{1}{8}$ | 0.125 | 3.175 |
| | $\frac{9}{32}$ | 0.2812 | 7.142 | 11 | | 0.116 | 2.946 |
| 2 | | 0.276 | 7.010 | | $\frac{7}{64}$ | 0.1093 | 2.776 |
| | $\frac{17}{64}$ | 0.2656 | 6.746 | 12 | | 0.104 | 2.642 |
| 3 | | 0.252 | 6.401 | | $\frac{3}{32}$ | 0.0937 | 2.380 |
| | $\frac{1}{4}$ | 0.250 | 6.350 | 13 | | 0.092 | 2.337 |
| | $\frac{15}{64}$ | 0.2343 | 5.951 | 14 | | 0.080 | 2.032 |
| 4 | | 0.232 | 5.893 | | $\frac{5}{64}$ | 0.0781 | 1.983 |
| | $\frac{7}{32}$ | 0.2187 | 5.554 | 15 | | 0.072 | 1.829 |
| 5 | | 0.212 | 5.385 | 16 | | 0.064 | 1.626 |
| | $\frac{13}{64}$ | 0.2031 | 5.158 | | $\frac{1}{16}$ | 0.0625 | 1.587 |
| 6 | | 0.192 | 4.877 | 17 | | 0.056 | 1.422 |
| | $\frac{5}{16}$ | 0.1875 | 4.762 | 18 | | 0.048 | 1.219 |
| 7 | | 0.176 | 4.470 | | $\frac{3}{64}$ | 0.0468 | 1.188 |
| | $\frac{11}{64}$ | 0.1718 | 4.363 | 19 | | 0.040 | 1.016 |
| 8 | | 0.160 | 4.064 | 20 | | 0.036 | 0.914 |
| | $\frac{5}{32}$ | 0.1562 | 3.967 | 21 | | 0.032 | 0.813 |
| 9 | | 0.144 | 3.658 | | $\frac{1}{32}$ | 0.0312 | 0.792 |
| | $\frac{9}{64}$ | 0.1406 | 3.571 | 22 | | 0.028 | 0.711 |

| Standard wire gauge | Fractions of inch | Decimal equivalents inch | Metric equivalents millimetres | Standard wire gauge | Fractions of inch | Decimal equivalents inch | Metric equivalents millimetres |
|---|---|---|---|---|---|---|---|
| 23 |  | 0.024 | 0.610 | 37 |  | 0.0068 | 0.1727 |
| 24 |  | 0.022 | 0.559 | 38 |  | 0.0060 | 0.1524 |
| 25 |  | 0.020 | 0.508 | 39 |  | 0.0052 | 0.1321 |
| 26 |  | 0.018 | 0.457 | 40 |  | 0.0048 | 0.1219 |
| 27 |  | 0.0164 | 0.4166 | 41 |  | 0.0044 | 0.1118 |
|  | $\frac{1}{64}$ | 0.0156 | 0.3968 | 42 |  | 0.0040 | 0.1016 |
| 28 |  | 0.0148 | 0.3759 |  | $\frac{1}{256}$ | 0.0039 | 0.0992 |
| 29 |  | 0.0136 | 0.3454 | 43 |  | 0.0036 | 0.0914 |
| 30 |  | 0.0124 | 0.3150 | 44 |  | 0.0032 | 0.0813 |
| 31 |  | 0.0116 | 0.2946 | 45 |  | 0.0028 | 0.0713 |
| 32 |  | 0.0108 | 0.2743 | 46 |  | 0.0024 | 0.0610 |
| 33 |  | 0.0100 | 0.2540 | 47 |  | 0.0020 | 0.0508 |
| 34 |  | 0.0092 | 0.2337 | 48 |  | 0.0016 | 0.0406 |
| 35 |  | 0.0084 | 0.2134 | 49 |  | 0.0012 | 0.0305 |
|  | $\frac{1}{128}$ | 0.0078 | 0.1984 | 50 |  | 0.0010 | 0.0254 |
| 36 |  | 0.0076 | 0.1930 |  |  |  |  |

# Appendix B
# Summary of the
# Chemistry of
# Photoresists †

### B.1 Liquid negative-working photoresists

Although well established photoresist systems such as dichromated gum arabic and iodoform sensitised Bakelite may still be found in use today in various branches of photomechanical engineering throughout the world, the systems listed below are thought to be used almost exclusively for PCM applications in Western Europe, North America and Japan.

*Polyvinyl (and other) cinnamates*

The Kodak KPR family of photoresists and Gemex PR1 are based on the polymer poly(vinyl cinnamate), whilst Tokyo Oka Kogyo OSR is based on poly(vinyl oxethylcinnamate) (figure B.1).

To formulate a liquid photoresist from the base polymer, solvents, sensitisers (to increase the photosensitivity of the polymer), additives (to prevent reactions occurring on contact with metal), thermal polymerisation inhibitors and non-photosensitive resins (such as Novolaks to improve adhesion to the metal and reduce swelling during development) are required.

On irradiation with ultraviolet light, cross-linking of the polymers occurs to produce a product of high molecular weight and low solubility in certain organic solvent mixtures (figure B.2). A photoresist developer is therefore formulated which dissolves the non-irradiated photoresist but which does not dissolve the irradiated, cross-linked material. Photoinsolubilisation is the cause of this class of photoresist (together with all the others in §3.5.1) being negative-working.

† References are contained in §3.8.

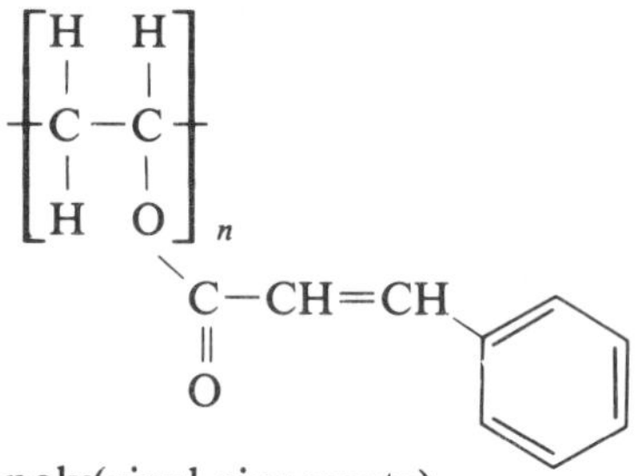

poly(vinyl cinnamate)

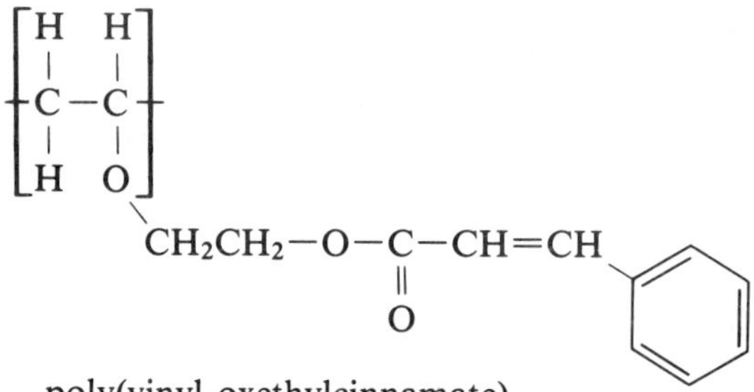

poly(vinyl oxethylcinnamate)

**Figure B.1**    Cinnamic acid derivatives.

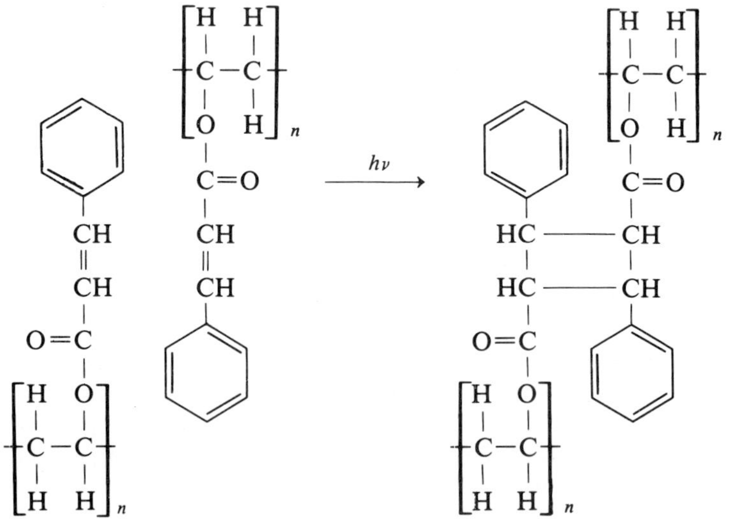

**Figure B.2**    Photochemistry of poly(vinyl cinnamate).

*Allyl ester resins*

Dynachem DCR 3140 (Robertsons PC 1000) comprises a solvent soluble 'pre-polymer' formed from diallyl phthalate (figure B.3) together with sensitisers and solvents. Irradiation of the photoresist causes the 'pre-polymer' to cross-link and form higher molecular weight polymers with greatly reduced solubility in organic solvents.

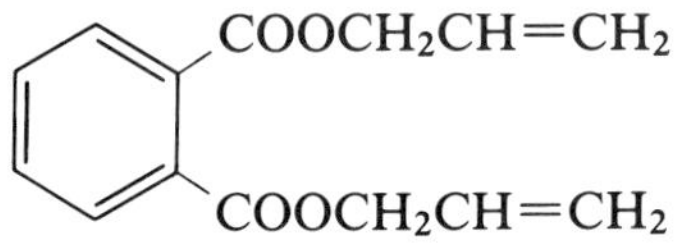
COOCH2CH=CH2
COOCH2CH=CH2

**Figure B.3** Diallyl phthalate.

*Cyclised rubbers*

More correctly termed partially cyclised polyisoprenes, cyclised rubbers have a complex structure, as shown in figure B.4, comprising isoprene, monocyclic and bicyclic structures. Bicyclic structures predominate to give the resist its desired chemical and physical characteristics.

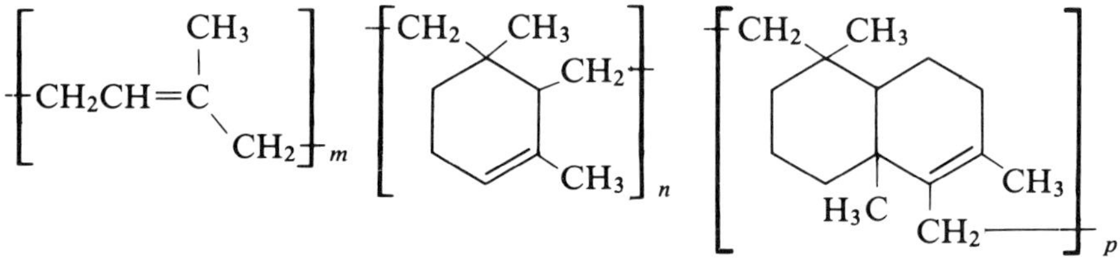
CH3
CH2CH=C
CH2
m
CH2
CH3
CH2
CH3
n
CH2
CH3
CH3
H3C
CH2
p

**Figure B.4** Structure of partially cyclised polyisoprenes (represented by ~~~ $CH_2$ ~~~).

The role of the sensitiser (usually a bis-azide) as shown in figure B.5 is especially important in the photochemical reaction which occurs on irradiation. It has been shown that one photon can cleave the two azide groups simultaneously to form a bis-nitrene. This intermediate quickly abstracts hydrogen atoms from two polymer chains and thus links them together with the sensitiser forming a bridge between the chains (figure B.6) (Shimizu and Bird 1977). The products are less soluble than the starting material in organic solvents such as xylene/white spirit (3 : 2v/v).

*Epoxy resins*

Photoresists in this classification comprise a photodimerisable epoxy resin, a hardener and solvents.

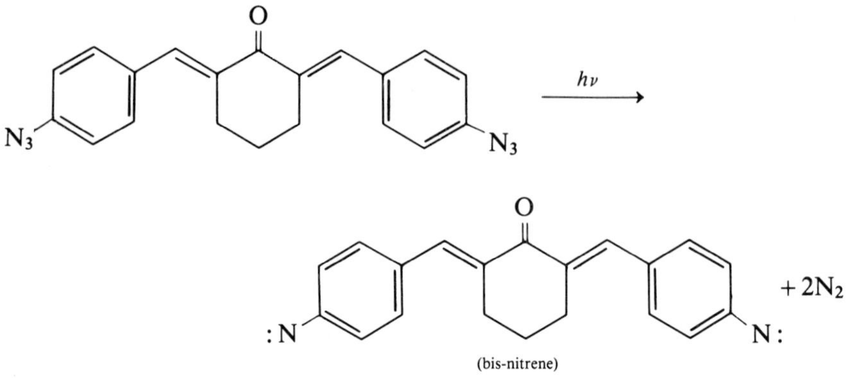

**Figure B.5**   Structure and photochemistry of a bis-azide.

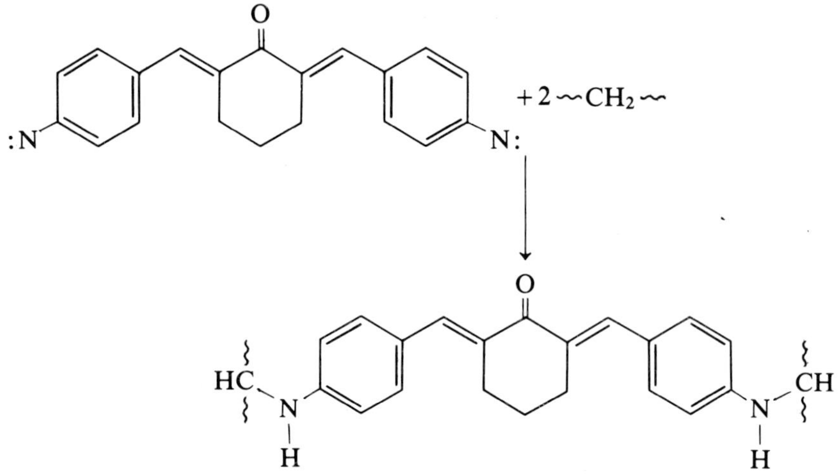

**Figure B.6**   Reaction of bis-nitrene with partially cyclised polyisoprenes.

The general structure of the resin is shown in figure B.7 where L is a light sensitive group, e.g.

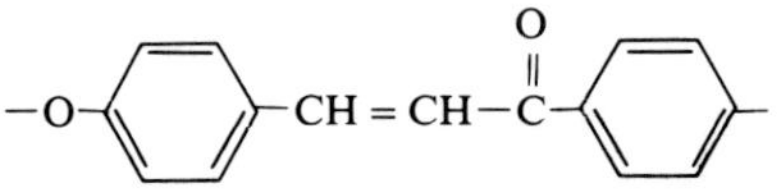

and B is a bis-phenol residue, e.g.

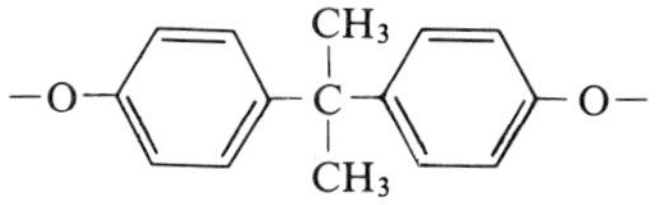

The resin is soluble in 2-ethoxyethanol (Cellosolve). On irradiation, the L groups cross-link and the resin becomes insoluble in the developing solvent (e.g. cyclohexanone/isopropanol).

After exposure, development, a low temperature bake (60 °C) and etching, coated substrates may be laminated to each other by heating to 135 °C in a press. The glycidyl ether (end) groups of the resin react with the hardener to effect further cross-linking and form an adhesive bond (see §3.6.6) (Rembold 1978).

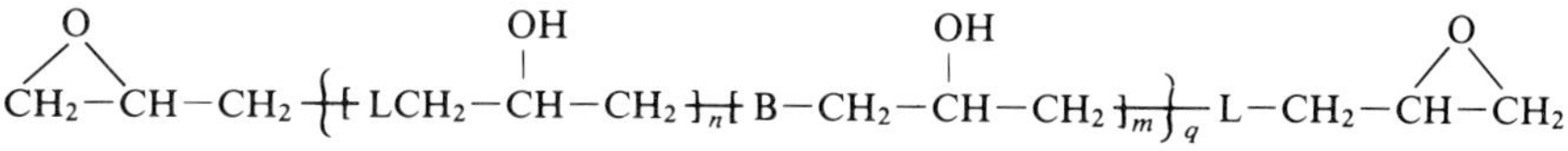

**Figure B.7** Structure of epoxy resin photoresist.

*Water soluble photoresists*

A few systems exist where the solvent is water and the main photoresist constituent is a synthetic polymer such as poly(vinyl alcohol) (PVA) or a natural colloid such as fish gelatin or casein. These systems are cheap, non-toxic and non-flammable, but must be sensitised by inorganic compounds. In the USA and Europe, Norland Products market Photoengraving Glue and NPR 29F photoresists based on a fish gelatin obtained from cod skins, whilst their NPR 40 formulation is based on PVA. PVA and casein photoresists are the most commonly used photoresists in Japan for PCM applications. The Japanese casein product is obtained from imported New Zealand milk.

The chemistry of these systems is not thoroughly understood, although they have been investigated many times over a long period of time. An outline of the chemistry is given below and processing details of commercial formulations are presented in table B.1.

Sensitisation of PVA, Photoengraving Glue and casein with ammonium dichromate produces a water insoluble product after irradiation. Ultraviolet light accelerates the reduction of $Cr^{VI}$ to $Cr^{III}$ and it is the latter species which reacts with the polymer or colloid to form an insoluble oxidation product. The sensitised photoresist has a limited pot-life dependent on pH.

NPR 29F is sensitised with an $Fe^{III}$ chelate. On reduction to $Fe^{II}$ by irradiation, followed by immersion in dilute hydrogen peroxide solution, a water insoluble product is formed which is hardened by a chemical treatment (see table B.1).

**Table B.1** Negative-working water soluble photoresists: processing details.

| | Dichromated photoresists | | | Fe$^{III}$ sensitised photoresist |
|---|---|---|---|---|
| Formulation | Norland NPR 40 | Norland Photoengraving Glue | Fuji Chemicals Resist no. 15 (FR-15) | Norland NPR 29F |
| Base | Poly(vinyl alcohol) | Fish gelatin | Casein | Fish gelatin |
| Pre-exposure baking | 65 °C for 10 min | 65 °C for 5–15 min | < 80 °C | 80 °C for 10 min |
| Chemical treatment after exposure | – | – | – | Dip in 0.1% hydrogen peroxide solution for 10 s |
| Development | Room temperature tap water | Room temperature tap water | 40 °C tap water | Room temperature tap water spray |
| Hardening treatment | (1) Dip in 8% chromic acid for 5–30 s (2) Rinse in water for 5–10 s (3) Dip in isopropanol for 5 s (4) Dry at 65 °C for 5 min | Not required | Optional dip in 5–10% chromic acid for 20–30 s | (1) Dip in 2% ammonium dichromate + 2% ferric sulphate solution for 15–20 s (2) Rinse in room temperature tap water (3) Dry at 50–60 °C |
| Post-development baking | 121–205 °C for 10 min | 260–288 °C for 5–10 min | Optional bake at 200–300 °C | 240–260 °C for 10 min |
| Strip | 5% Sodium hypochlorite at 82 °C for 1 min | 5–10% Sodium hydroxide at 71–82 °C for 30–60 s. See also table 3.6 | 10–20% Sodium hydroxide at > 60 °C | 10–30% sodium hydroxide at 75–80 °C (optionally with 10% v/v of 5% sodium hypochlorite for faster stripping) |

## B.2 Dry film negative-working photoresists

The original dry film photoresist introduced by Du Pont in 1968 was solvent developable. Ideally the developer should be non-toxic, non-flammable, easy to dispose of and cheap. For this reason semi-aqueous and ultimately totally aqueous developable photoresists were marketed and have proved themselves to be extremely useful products, even if more costly than their liquid counterparts.

*Solvent developable dry film photoresists*

Two types of photoresist in this classification include:
   (i) A photosensitive resin system where a photosensitive chemical structure (typically $-CH=CH_2$) is made part of a copolymer as shown in figure B.8. On irradiation the copolymer cross-links.

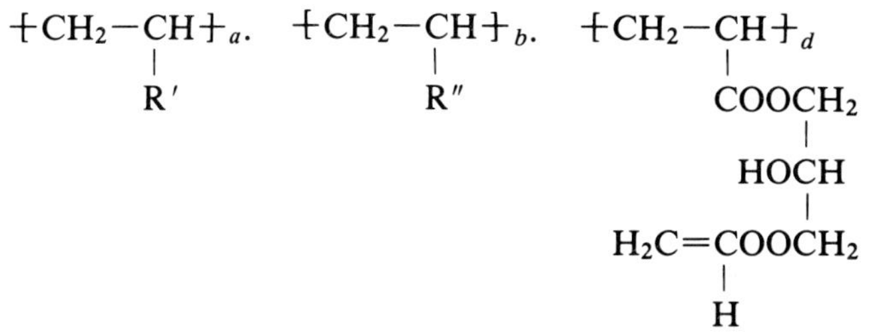

$$+CH_2-CH+_a. \quad +CH_2-CH+_b. \quad +CH_2-CH+_d$$
$$\begin{array}{ccc} | & | & | \\ R' & R'' & COOCH_2 \\ & & | \\ & & HOCH \\ & & | \\ & & H_2C=COOCH_2 \\ & & | \\ & & H \end{array}$$

**Figure B.8** Generalised structure of a photosensitive resin.

(ii) Photosensitive polyfunctional ester monomers of acrylic acid. These are formed by reacting together acrylic acid and polyfunctional alcohols as shown in figure B.9. Irradiation of the product leads to a polymer.
   Both systems need dyes, sensitisers and thermal polymerisation inhibitors in order to function efficiently.

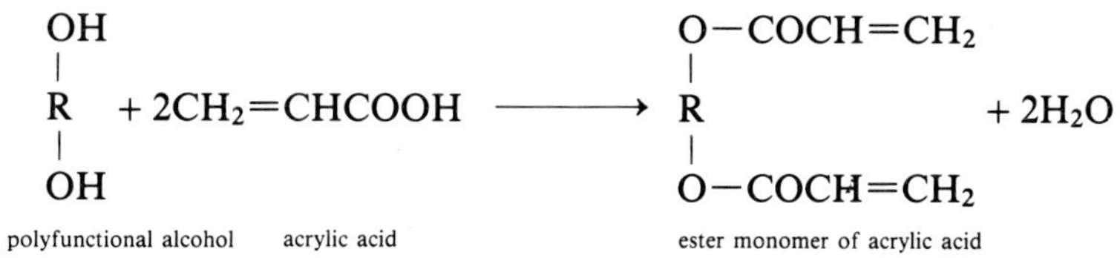

$$\begin{array}{ccc} OH & & O-COCH=CH_2 \\ | & & | \\ R \quad + 2CH_2=CHCOOH \longrightarrow & R & + 2H_2O \\ | & & | \\ OH & & O-COCH=CH_2 \end{array}$$

polyfunctional alcohol    acrylic acid                ester monomer of acrylic acid

**Figure B.9** Formation of a polyfunctional ester monomer of acrylic acid.

*Aqueous developable dry film photoresists*

A non-photosensitive copolymer of methacrylic acid and methyl methacrylate may be converted into a water soluble salt by reacting with an aqueous alkaline solution provided that a sufficient number of acid groups react to overcome the hydrophobic nature of the methyl methacrylate groups in the polymer.

If photosensitive polyfunctional ester monomers of acrylic acid are included in the resist formulation then free radical polymerisation will result in a hydrophobic product which will not be developed out after exposure and will therefore form the resist stencil on a substrate. This scheme is shown in figure B.10.

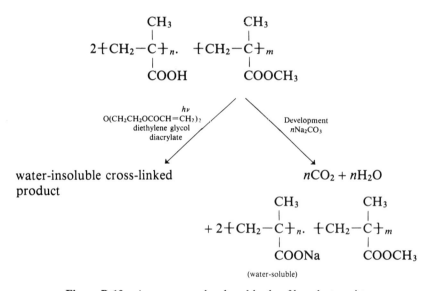

**Figure B.10** An aqueous developable dry film photoresist.

*Semi-aqueous developable dry-film photoresists*

The inclusion of some photosensitive water insoluble polyfunctional ester monomers of acrylic acid such as pentaerythritol triacrylate, $HOCH_2C(CH_2OCOCH=CH_2)_3$, necessitates additions of an organic solvent such as butyl Cellosolve to an aqueous developer in order to effect dissolution of unexposed photoresist. This developing system is then termed 'semi-aqueous'.

## B.3  Liquid positive-working photoresists

Positive-working photoresists are rarely used for PCM, being susceptible to chipping during spray etching (§3.7.5) and being incompatible with alkaline etchants. However, they are useful for manufacturing parts from thin (<0.025 mm) foils. In this instance undercut is sufficiently small that chipping does not occur. In addition the resist is very easily stripped from the delicate etched foils by rinsing in organic solvents.

*Ortho-quinone diazides*

Two types of photoresist are made commercially. Type A comprise:

(i) a non-photosensitive polymer such as a phenolformaldehyde Novolak resin;

(ii) a photosensitive component, called variously a sensitiser, photoactive compound (PAC) or inhibitor (as it functions to lower the solubility of the resist formulation). The sensitiser is usually a 1, 2-naphthoquinone diazide;

(iii) an organic solvent; and possibly

(iv) dyes, plasticisers and stabilisers.

Type B comprise:

(i) a photosensitive resin synthesised from very carefully controlled quantities of Novolak resin and a sulphonyl chloride derivative of 1,2-naphthoquinone diazide (Nakane *et al.* 1973);

(ii) an organic solvent; and possibly

(iii) dyes, plasticisers, stabilisers and a non-photosensitive polymer to improve physical characteristics of the resist and/or to lower formulation costs.

The chemistry of the photolysis of the photoresist proceeds as outlined in figure B.11.

The ketene is extremely reactive at room temperature and will react with traces of water in the coating to form 3-indene carboxylic acid which is alkali soluble (figure B.12).

If the coating is too dry for a reaction with water to occur, then the ketene will react with the phenolic groups of the Novolak resin instead. The latter product is not alkali soluble. For optimum processing it is recommended that, prior to exposure, photoresist-coated substances should be stored in the dark or under safelights in an atmosphere of 50% RH (relative humidity) at 21 °C for 15 minutes.

The developer consists of an alkaline solution and so the exposed coating is dissolved away (table B.2). The remaining resist stencil comprises the original coating. As this remains photosensitive, it may be re-imaged after etching using a different phototool and the substrate etched again! This technique may be useful when different features on the same surface need to be etched to different depths.

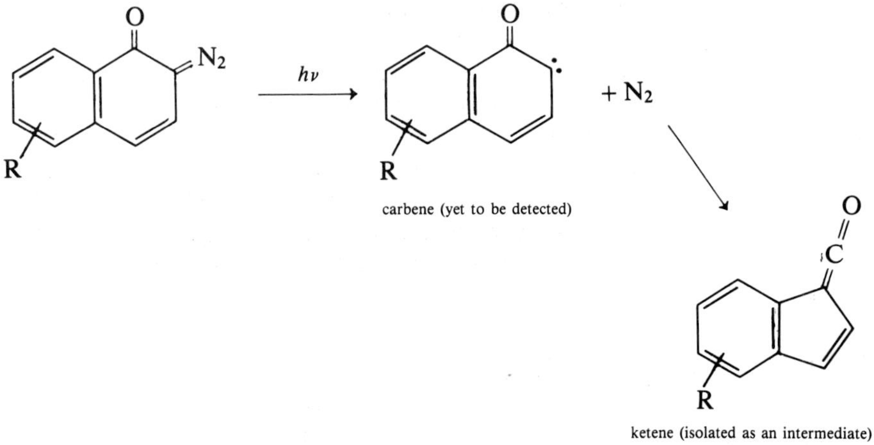
O
$N_2$
hν
O
$+ N_2$
R
R
carbene (yet to be detected)
O
C
R
ketene (isolated as an intermediate)

Type A; R = simple radical
Type B; R = polymer chain.

**Figure B.11** Photochemistry of naphthoquinone diazide derivatives.

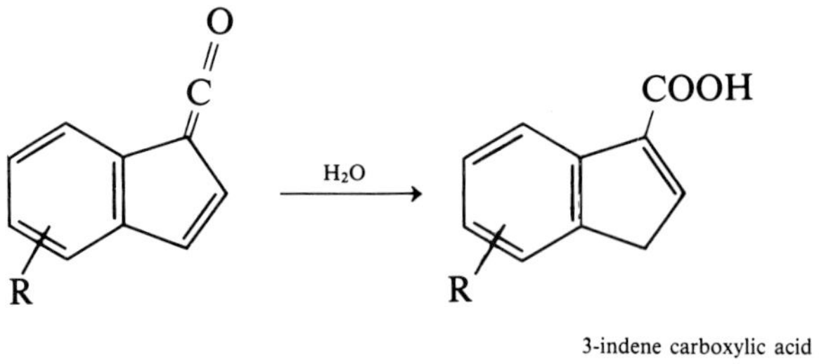
O
C
$H_2O$
COOH
R
R
3-indene carboxylic acid

**Figure B.12** Hydrolysis of indene ketene.

**Table B.2** Dissolution rates of positive-working photoresist constituents.

| Film constituent of Shipley AZ 1350J | Rate of dissolution of coating in Shipley AZ Developer ($nm\,s^{-1}$) |
|---|---|
| Resin only | 15 |
| Resin + sensitiser | 0.1–0.2 |
| Resin + sensitiser after exposure | 100–200 |

## B.4  Dry film positive-working photoresists

Kalle has patented a number of dry film systems. A typical coating comprises a polychloro compound, a Novolak resin as binder and a hydrolysable moiety such as an orthoester. Irradiation of the polychloro compound yields hydrochloric acid gas which dissolves in water present in the coating and hydrolyses the orthoester to a mixture of alcohols and esters. These products act as a cosolvent with the aqueous alkaline developer solution to effect dissolution of the Novolak resin (figure B.13).

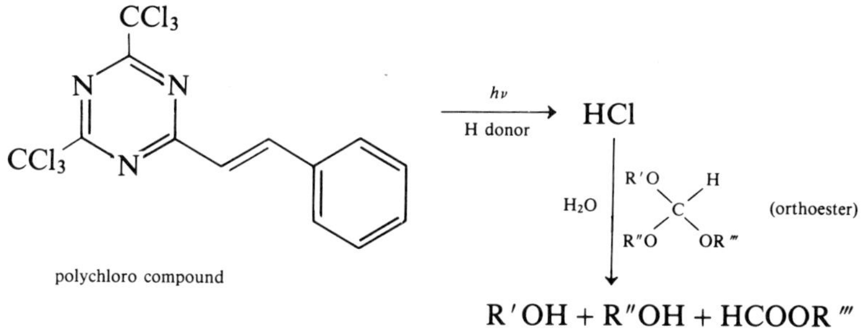

**Figure B.13**  Chemistry of a dry film positive-working photoresist.

# Appendix C
# Specific Gravity–Degree Baumé Conversions

Etchant densities are often given in terms of US heavy Baumé units (degrees) according to the formula:

$$\text{SG (specific gravity)} = \frac{145}{145 - {}^{\circ}\text{Bé}}$$

i.e.

$$ {}^{\circ}\text{Bé} = 145\left(\frac{\text{SG} - 1}{\text{SG}}\right). $$

Some equivalent values are tabled below:

| Specific Gravity | Degrees Baumé | Specific Gravity | Degrees Baumé |
|---|---|---|---|
| 1.00 | 0.00 | 1.15 | 18.91 |
| 1.01 | 1.44 | 1.16 | 20.00 |
| 1.02 | 2.84 | 1.17 | 21.07 |
| 1.03 | 4.22 | 1.18 | 22.12 |
| 1.04 | 5.58 | 1.19 | 23.15 |
| 1.05 | 6.91 | 1.20 | 24.17 |
| 1.06 | 8.21 | 1.21 | 25.16 |
| 1.07 | 9.49 | 1.22 | 26.15 |
| 1.08 | 10.74 | 1.23 | 27.11 |
| 1.09 | 11.97 | 1.24 | 28.06 |
| 1.10 | 13.18 | 1.25 | 29.00 |
| 1.11 | 14.37 | 1.26 | 29.92 |
| 1.12 | 15.54 | 1.27 | 30.83 |
| 1.13 | 16.68 | 1.28 | 31.72 |
| 1.14 | 17.81 | 1.29 | 32.60 |

*(continued)*

| Specific Gravity | Degrees Baumé | Specific Gravity | Degrees Baumé |
|---|---|---|---|
| 1.30 | 33.46 | 1.56 | 52.05 |
| 1.31 | 34.31 | 1.57 | 52.64 |
| 1.32 | 35.15 | 1.58 | 53.23 |
| 1.33 | 35.98 | 1.59 | 53.80 |
| 1.34 | 36.79 | 1.60 | 54.38 |
| 1.35 | 37.59 | 1.61 | 54.94 |
| 1.36 | 38.38 | 1.62 | 55.49 |
| 1.37 | 39.16 | 1.63 | 56.04 |
| 1.38 | 39.93 | 1.64 | 56.58 |
| 1.39 | 40.68 | 1.65 | 57.12 |
| 1.40 | 41.43 | 1.66 | 57.65 |
| 1.41 | 42.16 | 1.67 | 58.17 |
| 1.42 | 42.89 | 1.68 | 58.69 |
| 1.43 | 43.60 | 1.69 | 59.20 |
| 1.44 | 44.31 | 1.70 | 59.71 |
| 1.45 | 45.00 | 1.71 | 60.20 |
| 1.46 | 45.68 | 1.72 | 60.70 |
| 1.47 | 46.36 | 1.73 | 61.18 |
| 1.48 | 47.03 | 1.74 | 61.67 |
| 1.49 | 47.68 | 1.75 | 62.14 |
| 1.50 | 48.33 | 1.76 | 62.61 |
| 1.51 | 48.97 | 1.77 | 63.08 |
| 1.52 | 49.60 | 1.78 | 63.54 |
| 1.53 | 50.23 | 1.79 | 63.99 |
| 1.54 | 50.84 | 1.80 | 64.44 |
| 1.55 | 51.45 | | |

Footnote: On rare occasions the density term 'degrees Twaddell' ($^{\circ}$Tw) is encountered in etchant formulations. It relates to SG as follows:

$$(^{\circ}Tw) \times 5 + 1000 = 1000 \times SG$$

# Index